The Sky at Night

BOOK OF THE

MOON

The Sky at Night

BOOK OF THE MOON

A Guide to Our Closest Neighbour

Dr. Maggie Aderin-Pocock

10

BBC Books, an imprint of Ebury Publishing
20 Vauxhall Bridge Road,
London SW1V 2SA

BBC Books is part of the Penguin Random House group of companies
whose addresses can be found at global.penguinrandomhouse.com

Illustrations all by Greg Stevenson apart from page 50, 145, 168 (Jonathan Baker); 163, 165 (Royal Astronomical Society of Canada); 79 (Marshack, A. 1970); 76 (Pasildes); 75 (Philip J. Stooke).

Translation credits: *Somnium* by Tom Metcalfe, Herodotus by A.D. Godley and *Midnight Poem* by H.G. Wharton.

Endpapers and Moon drawing designed by Maria Rikteryte.

This book is published to accompany the television series entitled *The Sky at Night* first broadcast on BBC One in 1957.

First published by BBC Books in 2018.

www.penguin.co.uk

A CIP catalogue record for this book is available from the British Library.

ISBN 9781785943515

Typeset in 11/15 pt Garamond MT Std
by Integra Software Services Pvt. Ltd, Pondicherry

Printed and bound in Great Britain by Clays Ltd, Elcograf S.p.A.

Penguin Random House is committed to a sustainable future for our business, our readers and our planet. This book is made from Forest Stewardship Council® certified paper.

To family and friends that have patiently known me
so long, I dedicate this to you. Starting in no particular order
(well, actually, chronological):

To my Mum for carrying me, inspiring me
and making me stronger.
To my Dad for carrying me to higher planes,
helping me to think and giving me the 'moon on a stick'.
To my sisters, Sue and Hal, for generating in me a passion
for music, drama and an inquiring mind.
To my sister Grace who taught me to truly care
for someone other than myself.
To Martin – my co-conspirator in life – for giving me
stability and understanding (especially when
writing gets frustrating).
To Lori, I carried you, but now you carry me to a
better place on higher planes, you make me stronger,
and you make me want to know more so I can share it
with you. You enable me to see myself better, in the
dazzling mirror that is you. But most of all, for giving me
the best tangible reason for wanting to make the world
a better place for everyone, but especially for you.

To all my fellow lunatics out there, may we all,
each and every one of us, get closer to our goal.

CONTENTS

INTRODUCTION:

WHY BE A LUNATIC?

I ONCE HAD a friend tell me that they thought of me when they see the Moon. It was an innocent comment but to me there could be no greater compliment. The Moon is my yin and yang. It has driven my career and influenced how I have lived my life. In many ways it has defined what I have become, so for me to be associated with it in that way is wonderful.

We stand at an amazing juncture in history. Space and astronomy are truly having their moment, with new discoveries and developments happening all the time – from the first detection of gravitational waves* to the discovery of previously unknown exoplanets (planets orbiting the distant stars we see out in the night sky).

As technology develops it gets easier to look further and further afield. Yet sometimes looking at what we can see in our own backyard can be just as satisfying, especially because what we discover locally is more accessible to investigate.

This is definitely the case with our local companion, the Moon. Dominating our night sky, it is easily visible with the naked eye and we have been studying it ever since we were first able to look up and wonder. And yet, for all our observations and investigations, the Moon manages to hold on to some of its mystery. We seem to know surprisingly little about it. It is still uncertain exactly where it came from and how it was formed, yet we continue to theorise and speculate. In this book, I want us to go on a journey of

---

* Gravitational waves are 'ripples' in the fabric of space-time caused by some of the most violent and energetic processes in the universe. Albert Einstein predicted their existence in 1916 in his general theory of relativity.

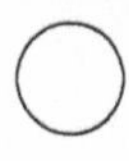

understanding: exploring what we know and debating what is yet to be discovered.

As we celebrate the fiftieth anniversary of the Moon landings in 2019, let us develop a better awareness of the incredible life of the Moon. We will travel through its long and fascinating past and learn how it supports our fragile Earth in the present, and we will look forward too, to a relatively unknown but surely exciting future that awaits us regarding our ever-changing relationship with the Moon.

But most of all, and most importantly, let us celebrate our celestial partner and bring out the inner lunatic that lies within all of us.

MY FATHER, THE MOON AND ME

Ever since I can remember, the Moon has played a vital role in my life and I've always been extremely happy to describe myself as a self-certified lunatic. I love the Moon; it fascinates me and, as far as I can tell, I've always had that fascination since I was very small. This might not be surprising as I was born in 1968 and the Moon landing happened the following year. Indeed, I was taking my first small steps while Neil Armstrong was taking his giant leap for mankind. Having been brought up in that era of excitement and exploration, it was perhaps inevitable that I have always looked towards the Moon. In many ways, I feel that the Moon has always been there with me, leading me along my path and steering me into my career as a space scientist.

I cannot recollect my first memories of the Moon, but I

believe that my formal introduction came via my father, who would tell me tales of his childhood in Nigeria. When he was growing up he owned a Raleigh bicycle. Now, Raleigh was the Rolls-Royce of bicycles, as far as my father was concerned. And my father would ride his beloved Raleigh bicycle a long way to get to school, some 12 miles or more, which meant he had to start his day early and always returned home late. But there were no roads or street lights to guide his way. Instead, he cycled across the sand and dust of the savannah with just the Moon lighting his way. He told us of its huge size and astounding beauty, how he saw it as a companion on his long journeys home. I knew if the Moon was a good enough companion for my father, then it was definitely good enough for me, growing up in north London.

So the Moon became a reassuring presence in the night sky, and these feelings would only intensify as I struggled to sleep throughout my childhood, battling insomnia. I would lie awake in bed for hours, waiting for sleep to come. Everyone else would be deep in the land of nod, but somehow sleep would elude me. After a while, giving up on pretending, I would climb out of bed and enter a world of silent darkness. Night time can be a sinister realm for a child. Nothing is quite as it seems during the day and the most innocent objects take on a menacing aspect. Moving about the house, I would try to create as little disturbance as possible as the other members of my family slept, while tiptoeing to a window where I could see the sky clearly. Drawing back the curtains, I would reveal the dazzling brightness of moonlight streaming in across the room, banishing the world of darkness and creating a domain of wondrous light. Just viewing this spectacle for a few minutes would calm my nerves and relax me enough to attempt to sleep again. As with my father, the Moon was my friend and companion.

My father would go on to say that he missed the African Moon of his youth. That it didn't quite feel or look the same when viewed through the cloudy London skies. Although I could sympathise with his point of view, I felt that he was missing a certain beauty that only an old grey city can have when bathed in moonlight.

For instance, I remember one evening walking across Hampstead Heath on my way home from school. Growing up in London, I didn't get much chance to see many stars. Cloudy skies and light pollution rather limited what was available. But the Moon was something else. Our school was situated in Highgate and at that time we lived in a council flat in Belsize Park. The quickest route between the two was through Hampstead Heath. We didn't walk far into the Heath, as we were just kids, but there was a shortcut home along a path – and although we could take the street route, it took much longer and was not half as pleasant. That evening, my younger sister Gracie and I were walking home, and it must have been winter because it was dark despite being only 5pm or so in the evening – but I don't remember being cold. I just remember this huge shining silver disc in the sky, light pouring off it so strongly that we could even see our moon shadows on the ground. I was mesmerised by it.

Another evening, when I was 12 or 13, I was walking with my sister (Gracie again), either going to the shops or perhaps on our way back home. As always, I looked up and observed the familiar presence in the night sky but, this time, something was wrong. I had seen the Moon the night before and it had been nearly full, and now when I looked at it, it was far from full, its phase was wrong. I had been looking forward to the full moon and now it was a crescent, with much of its surface rendered invisible. Where had that section of the Moon gone? I started to doubt my memory. There are few things you can rely on but the phases of the Moon are one of them. They are, and always have been, predictable, like clockwork. This didn't make sense and gave me a real sense of foreboding.

I later found out that the Moon that night was in partial eclipse and I was thrilled that I had been able to observe such an event. Though I was young, I already had an inherent awareness of the Moon and its phases, and what I might expect to see in the sky. This was in the days before the internet and I just had books to rely on, so I really had to be interested to find that sort of thing out.

We may not have had the internet but we did have a television (yes, I am not *that* old), and as I grew up my interest in the Moon was enhanced by watching *The Sky at Night*. My parents would give me special permission to stay up late and watch the programme, knowing how much I loved it. Suddenly, there was more to see than just the Moon out there. With Patrick Moore's guidance, I was now able to spot constellations and planets, and I was hearing about missions investigating the solar system and telescopes helping us to understand the universe. I was hooked and wanted to know more.

I started looking at books on astronomy in the library and then, the next step, I decided to save up my pocket money to buy my own telescope. But this was going to be a tall order. I scanned magazines, looking at the systems available but, sadly, they were all beyond my teenage budget. I decided to start somewhere and bought a small, cheap telescope from Argos – it was just a little plastic one with lenses and it wasn't very good. It suffered from something called chromatic aberration, which is when the light coming through the lenses gets split up, so you almost see three separate images of different colours. I'd have been better off buying a pair of binoculars; I know that now.

Because of this, I thought that my observations would remain limited to the naked eye but a few weeks later I found a notice in a local magazine advertising telescope-making classes. It was part of Camden's adult education programme. I was stunned; was it possible for me to make a quality telescope? I went along to the class and found that not only was I the youngest there at 15, I was also the only female – something that would happen plenty of times throughout my career, though happily more women are taking up careers in science nowadays. But I was with like-minded people. I started to make a 150mm Cassegrain system (see page 10); quite challenging for a novice. Some of my companions at the class were there for the technical challenge of making the telescope, but, for me, it was a means of getting closer to the Moon and stars.

We were actually making the telescope's mirror, the heart of any reflecting telescope system, by grinding and polishing thick slabs of glass, which when shaped would be covered in aluminium to make a shiny, reflective surface.

If you put two of these glass slabs together with some abrasive powder between them and then rub them together, the one on the bottom becomes concave (the middle of the flat surface is worn away so a well is formed), and the top one becomes convex (the edges are worn down so the middle ends up being raised, like a hill). You do this for months (I used to grind my mirror watching *Star Trek*) using finer and finer abrasive powders, until you have one concave and one convex spherical surface with a very smooth finish. You work at the concave surface at this stage and you gradually change its shape from spherical into something called a parabola – more like an open U-shape. After adding the reflective coating, this becomes your primary mirror, and it can be mounted in a box with other simple mirrors to form your telescope. It takes months and lots of careful measurement to judge the shape, but it's a really good, cheap way of making your own telescope. And the real joy for me was that this was something that I had made with my own hands and it got me closer to the Moon.

TYPES OF TELESCOPE

There are three main types of telescope: the refracting, the reflecting and the catadioptric telescope. I will give each type a quick mention here.

Refracting Telescopes

This was the first type of telescope that was invented and it was this kind that was used by Galileo to make his early observations of the Moon. The light passes through a series of lenses that magnify the image and then focus it through an eyepiece into the back of the eye (the retina). The challenge with these telescopes is that the quality of the lens needs to be good. The glass that was available to Galileo in the seventeenth century probably had small air bubbles and impurities in it. As the light passes through the lens, these imperfections have a detrimental effect on the images that are produced.

Better quality glass is available today and plastic lenses are now often used in cheaper telescopes. These plastic lenses have the advantage of being lightweight and the plastic will be superior to the glass that was used by Galileo, but at the very cheap end these lenses can still suffer from problems like the chromatic aberration that my first telescope had.

Reflecting Telescopes

This type of telescope was first built by Sir Isaac Newton in 1668, although the principles of the design were around for a number of years before. It uses a reflecting surface rather than lenses to focus the light, which alleviates the problems associated with transmitting light through poor-quality glass. Reflecting telescopes come in many different forms, the most simple of which is the aptly named Newtonian telescope, which consists of a mirrored parabolic surface (the open U-shape surface mentioned earlier, which can focus light gathered from far away) and a flat mirror mounted in a box. This type is popular in telescope-making classes and has a characteristic look, as you do your observing from the front end of the telescope, but off to the side so you don't block the view.

There are many other variants of the reflecting telescope including the one I made as a child, a Cassegrain telescope, which also has a parabolic mirror but, in this case, the flat surface is replaced by a shaped one (a hyperbolic shape, for any who remember their conical sections at school) that bounces the light gathered back down through a hole in the main mirror. This leads to a smaller, more compact telescope – but it is harder to make, especially for a young novice. Many of the large professional telescopes, including those in the 8m class and bigger, use the Cassegrain configuration.

As an aside, I have actually held one of Newton's telescopes that he made and presented to the Royal Society in 1672. I was participating in a photoshoot at the Society and saw a strange object sitting on one of the library shelves. It was a beautiful thing but it was in the way, so I grabbed it to relocate it to somewhere else. As I held it someone exclaimed, 'Hey, watch it, that was Newton's second telescope!' My immediate response was panic and I almost dropped it, but sanity reigned and I was able to put it down intact and out of the way. An amazing piece of history saved for future generations.

Catadioptric Telescopes

The third type of telescope is the catadioptric telescope, which combines a reflecting telescope with lenses. The most popular form of this telescope is the 'Schmidt–Cassegrain'; many high-quality amateur telescopes are of this design. Manufacturers like this system because it can be made using simple spherical optics (the easiest shapes to machine manufacture or make by hand) and can be housed in a compact portable telescope, which makes it very popular.

While we were working on our mirrors, we'd speak about astronomy and share our enthusiasm, discussing what we might see through our homemade telescopes once they were completed. And after the class, if it was dark enough, we would go to the observatory on Hampstead Heath and look at the night sky using the more powerful telescopes there. The class is still going, and they email me every so often. It was wonderful to find a community of people who shared my passion.

My telescope developed well and I thought about building a tracking system for it. I did this as my first-year project while studying for my physics degree at Imperial College in

London. I got the telescope bug so badly that I started to specialise in optics too, going on to use it in my PhD. Finally, I managed to work on one of the largest telescopes in the world, Gemini South in Chile, an 8.1m triumph of engineering for which my team and I built a high-resolution spectrograph.* It was a real dream come true and this project led on to my career in space science, working on other instruments including some that were used on the James Webb Space Telescope, one of the world's largest space telescopes currently being built by NASA and ESA (European Space Agency). But it all began from that early passion and fascination I had for the wondrous sky at night and, of course, the Moon.

BEYOND BORDERS

If it was my father who introduced me to the Moon, it was Oliver Postgate who gave me the early desire to travel there. As a three-year-old I watched *The Clangers* with an almost religious

* A spectrograph is an instrument that splits light up into its component colours to allow the observer to analyse it. From this analysis many different things can be inferred, such as the motion of the object and its chemical composition.

conviction. My favourite part was the introduction, which varied from episode to episode but always included Postgate's gentle, melodious voice telling me of the possible delights that lay out there, beyond our planet... My one desire in life was to get in a rocket and travel outward – stopping off at the Moon and then travelling onward to the world of the Clangers.

I have always seen *The Clangers* as a bit of a gateway drug that leads to the world of science fiction, a genre that I grew to love in my youth. As a child, I suffered severely from dyslexia, and though I have now grown to know and love this in myself, it was a real barrier when I was young. When you first go to school, everything is hinged on the ability to read and write, and as a dyslexic this was something that I found very hard, especially because this was 1970s Britain and my condition remained undiagnosed. I was just considered dumb and dumped into the school's remedial class. There I sat with my desire to travel into space feeling like a pipe dream. Yet it was my parents who told me that if I worked hard enough, anything was possible, and who encouraged me to look for stories that interested and excited me. This is where I found science fiction ... and suddenly reading was worth the effort.

My desire to travel to the Moon therefore grew out of my father's stories, my love of *The Clangers* and my obsession with science fiction, yet it was also fuelled by a feeling of not really belonging anywhere. As a black child growing up in London during the seventies, I did not really feel very British. My colour seemed to be a barrier to my fellow students who would often tell me to go back home where I belonged, but on the other hand my relatives would call me a lost Nigerian as I had never been to the country and did not speak the language. Using the very handy 'retrospectroscope', I realise now that space had a strong appeal to me because it transcended national barriers. Viewed from space we are one.

The programme that epitomised these feelings for me in so many ways was *Star Trek*. In the original series, there were people from all over the world having adventures on the Starship *Enterprise*. Here we had reached a time in the future where national barriers did not exist. We were part of the United Federation of Planets, together with many different alien lifeforms. I wanted to be Spock – logical, unemotional and a whizz at science. Yet for me the star of the show was the one and only Lieutenant Uhura, the communications officer. She was my ultimate role model. Here was an accomplished black woman, a critical part of the team, gracing our screens. During the 1970s there were not many black people on TV so to have Uhura there doing all the things that I had dreamed about all my life really made an impression. And that impression has stayed with me. So much so that when I was invited to sit on a panel to celebrate the fiftieth anniversary of the first episode of *Star Trek*, I jumped at the chance. It was not until later that I found out that the actress who played Uhura, Nichelle Nichols, was going to be sitting on the panel too. I like to think of myself as a professional when it comes to public speaking, but in this case I was beside myself with joy. There, sitting on a stage next to me, was an amazing icon and one of my all-time heroes.

The sentiment of the human race living in harmony is a subject that is often discussed, but I think it was best summed up by Apollo 11 astronaut Michael Collins. I like to give him a special mention as he is the one that we hear the least about as he stayed on board the lunar orbiter while Neil Armstrong and Buzz Aldrin landed on the Moon's surface. In an interview he was asked about his strongest memory of the Apollo 11 mission. His response was:

> I really believe that if the political leaders of the world could see their planet from a distance of 100,000 miles their outlook could be fundamentally changed. That all-important border would be invisible, that noisy argument silenced. The tiny globe would continue to turn, serenely ignoring its subdivisions, presenting a unified façade that would cry out for unified understanding, for homogenous treatment. The Earth must become as it appears: blue and white, not capitalist or Communist; blue and white, not rich or poor; blue and white, not envious or envied. Small, shiny, serene, blue and white, fragile.

My desire is still to travel out in space and see what Michael Collins saw, but I think that we should all look to the Moon and space to get a feel of what we could become if we put our preconceived barriers to one side and live as one people, the human race.

AN ICONIC INFLUENCE

Finally, I'd like to make a special mention of another person who created a deep impression on me while I was growing up. If *The Clangers* and *Star Trek* first piqued my interest in the fictional mysteries of outer space, it was the presenter of another TV programme who helped me to develop a deeper understanding and appreciation of the genuine wonders of the universe.

I have already mentioned I was an avid watcher of *The Sky at Night.* Its legacy as a British institution is wide-ranging and deep. It is one of the programmes that many generations of people remember watching and it has that transgenerational appeal because it has been on our screens for so long (it has reached a pensionable age now as it is in its early sixties). But what made the programme most iconic was its presenter, Sir Patrick Moore.

Patrick was a self-taught astronomer with a phenomenal passion for the subject that imbued every book, television and radio programme he did. As the author of over 100 books and the presenter of *The Sky at Night*, his influence was enormous.

Patrick had a life-long love affair with the subject of the Moon; in fact, his first book in 1953 was entitled *A Guide to the Moon.* Through hours of dedicated observations, he became a highly respected lunar observer and was soon picked out by the then director of the lunar section of the British Astronomical Association, a Mr H.P. Wilkins. Patrick had a particular talent for spotting features on the cusp of the far side of the Moon via a process called libration. Although only one side of the Moon faces Earth, it is in fact possible to see more than 50 per cent of the Moon's surface (see pages 55–7 for more on this). Moore and Wilkins produced a number of high-resolution lunar maps published in 1946 and 1951, with a further revision in 1961 after Wilkins's death.

The fact that Patrick was observing the Moon at this time was fortuitous because during this period many lunar missions were being planned. Patrick was very proud of the fact that the Russians used the maps he had developed with Wilkins to prepare their photographs of the Moon from space.

The birth of the space era was also fortuitous for Patrick's television career. On 24 April 1957, at 10.30pm, Moore presented the first episode of *The Sky at Night*. With the launch of Sputnik later that year, the BBC felt that a regular programme highlighting the exciting developments taking place was necessary and *The Sky at Night* has been running ever since.

During an epic, world-record-breaking period of 57 years, Patrick presented over 700 episodes, only missing one due to a severe case of food poisoning. Patrick died in 2012 and I never met him – though that might have been for the best as his ideas on immigration and women within the BBC were at opposite poles to mine! But if we had met I would have hoped that our shared love for the Moon and all the night sky has to offer would have transcended the politics and left us as kindred spirits united by a love of the cosmos and everything out there.

It is my great honour now to be one of the presenters of *The Sky at Night* myself and I hope that on that programme, and in this book, I can inspire you with some of my enthusiasm in the same way that Patrick Moore did for me so many years ago.

On our journey we will look back into the depths of history and reflect upon the ways in which the Moon has influenced and inspired humankind across time, cultures and countries. We will also learn more about our mysterious neighbour in the present day – how it protects our precious Earth and is a key player in the story of our planet. Then, we will propel ourselves forward in time and look toward the possible future of the Moon – one where man will return (and women will make their first footsteps on its surface), space tourism will be widely available and colonies might settle there. But before we begin our voyage to the Moon, let's first familiarise ourselves with the basics . . .

MOON 101:

THE BASICS

THE MOON has fascinated humankind since the beginning of history and has long captured our imagination. After years of investigation and scientific progress, we now know much more about the science of the Moon – such as what it is made of and, at least in theory, how it was formed.

So let's begin by getting to grips with the scientific fundamentals of the Moon. The term '101' is used in some universities and colleges around the world to indicate that a course is an introduction to the subject and requires no previous knowledge, which is precisely the case here. There will be times when the science gets complex but I can assure you, by the end of this chapter, you'll have a firm foundation for everything you need to know. And what better place to start than with its name?

WHY IS THE MOON CALLED THE MOON?

It seems like an odd question; what else would it be called? But in the past our partner through the solar system has had a number of different names. To the Romans it was Luna (or 'lunar' in English); she was the goddess that personified the Moon and this term is still used today. The Greek version, Selena, was the name of their Moon goddess. You still hear it around as a girl's name although it is less associated with the Moon these days.

It might come as a surprise but giving the Moon its official name was actually one of the first things that was done by the International Astronomical Union (IAU) when it came into existence in 1919. The members did this because they wanted 'to standardise the multiple, confusing systems of nomenclature for the Moon that were then in use'.

The reason why they went for the somewhat basic name 'Moon' rather than something more exotic was because the name had already been in use for millennia, and in a range of different languages. Given the IAU was a newly formed organisation, it probably seemed like a good idea not to rock the boat too much on its first outing.

But, of course, it's a slightly confusing choice because ours is not the only moon. In the past, as we looked out into the solar system, we realised that there were many planets with moons orbiting them. In fact, we are still discovering more moons out there as we explore further into space. Although they are, fundamentally, all just called moons, we came up with more interesting naming systems for these satellites, often related to the name of the planet they orbit.

The planet Jupiter, for example, is named after the head honcho of the Roman deities. Jupiter was the god of sky and thunder and by all accounts a bit of a lad. Jupiter's largest moons – Io, Europa, Ganymede and Callisto – are named after his Greek counterpart Zeus's sexual conquests, which is one way of achieving immortality, I suppose. But names have been given to just 53 of Jupiter's 69 moons.

For the planet Mars, named after the Roman god of war, things were kept simple: it only has two moons so they were named after Deimos and Phobos, the sons of the Greek equivalent of the god of war, Ares.

The planet Saturn, named after the Roman god of agriculture, has many, varied moons. To date, 62 have been confirmed but only 53 have been named. The Greek theme continued here with the moons being named after Greek mythological figures. But by the time moons were spotted around Uranus the classical naming system had been dropped and characters from Shakespeare's plays and a poem by Alexander Pope were chosen instead.

Interestingly, the word 'moon' seems to stem from an old English word derived from the Germanic word *menon*, which in turn is thought to come from an Indo-European word, *menses*, meaning 'month' or 'moon'. So Moon seems like an appropriate term that has origins from across the world. Well, in English, it works for me.

THE MOON'S PHYSICAL CHARACTERISTICS

The Moon is an almost spherical lump of rock, gravitationally tethered to the Earth with an elliptical orbit (which means its path around the Earth is oval-shaped rather than a round circle). I say an *almost* spherical object because the Moon, like the Earth and many of the planets, is oblated, which means it is a slightly squashed sphere with the pole-to-pole distance being shorter (in this case only about 2km shorter) than the equatorial diameter.

The table below sets out some facts about the Moon to get us started:

Parameter (unit)	Value
Average distance to Earth (km)	385,000 (239,000 miles)
Axial tilt (degrees)	1.5
Rotational period (Earth days)	27.32
Orbital period (Earth days)	27.32
Maximum Temp (°C)	120 (248°F)
Minimum Temp (°C)	–247 (–413°F)

These facts are all very well and give us a good starting point for understanding the Moon. Yet I think it is easier to describe its characteristics relative to something with which we're already very familiar: the planet we live on. So the table below highlights some other characteristics of the Moon and gives an indication of how these key measurements compare to the Earth:

Parameter (unit)	Value	Compared to Earth
Average diameter (km)	3,474 (2,159 miles)	27%
Volume (km^3)	21.9 billion	2%
Surface area (km^2)	37.9 million	7.4%
Mass (kg)	7.35×10^{22}	1.2%
Density (kg/m^3)	3,344	60%
Gravity (m/s^2)	1.6	16.7%

Now let's look at some of these characteristics in more detail.

Size Matters

Compared with the Earth, the Moon is actually very small, having around a quarter of the diameter of our planet. This small diameter means that the Earth has a volume nearly 50 times that of the Moon.

The surface area of the Moon is also surprisingly small, just 7 per cent of the area that the Earth has. This means that the continent of Asia, which has a surface area of 44.4 million km^2, is actually larger than the Moon's surface. When you also take into account that only around 30 per cent of the Earth's surface is land, the Moon's surface area really is pint-sized compared with the Earth.

Although all these statistics make the Moon sound tiny, no other moon in our solar system is bigger in comparison to the size of the planet it orbits. Its large size and relatively close proximity means that it has a strong influence on our planet in many different ways (see page 108).

Weighing It Up

As the Moon is so much smaller in volume than the Earth, it will come as no surprise that its mass is, as expected, minor compared with that of the Earth, weighing in at just 1.2 per cent of our planet's mass. But even though the Moon is significantly smaller than the Earth, it has a curious property: it is lighter than it should be if both bodies were made out of the same material. In fact, the Moon's density is just 60 per cent of that of our planet, even though the two bodies have a similar chemical composition. This is because the Moon's internal structure is significantly different to that of the Earth's (see pages 30–3). That said, the Moon is no lightweight: it's actually

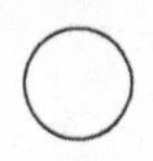

the second densest moon in the solar system, coming in a close second to Io, one of Jupiter's moons.

THE LUNAR LANDSCAPE

The surface of the Moon is rather lumpy, with its top elevations about 8km higher than the mean level of the surface, and lowest depths about 9km below the mean. This closely matches the range that we have on Earth from the highest mountains to the lowest part of the sea floor, but on the Moon this range is on a much smaller body.

On Earth, this 'dynamic range' in topography is caused by plate tectonics. As the plates collide they throw up mountain ranges, and, as other plates are forced below each other and sink down, trenches are formed. On the Moon, however, there is a very different process at work. Here the range in elevation is due to the craters that have been formed on the Moon's surface over billions of years. These craters vary in depth from hundreds of metres (the deepest crater on the Moon is the Aitken basin, which is about 12km (7.5 miles) deep from its raised rim), to micrometres, i.e. one-thousandth of a millimetre. And the Moon's surface is absolutely covered in them.

What's more, the Moon's atmosphere is so thin (see page 43) that very little erosion takes place. In fact, the rate of erosion is just 1cm every 20 million years. So craters that were formed billions of years ago can still be evident today. Indeed, a crater will stay on the lunar surface virtually for ever unless its presence is eroded by the arrival of fresh impact craters.

Luckily for us, on a clear, still night it is possible to see quite a bit of detail on the Moon's surface, if its phase is right for observation (see pages 139–42 for an explanation of phases). Bright and dim areas can be seen and its pocked and cratered surface can be observed with just the naked eye or, more clearly, with binoculars or a telescope. The dark areas are called maria (pronounced '*mar*-ee-a') and the light areas are called highlands or terrae, and they mark the two distinct types of the lunar terrain. Now we know how to spot them, let's look at these in more detail.

The Highlands

It is thought that the lunar crust was created about 4.5 billion years ago, soon after the Moon's formation, out of a sea of lava called a 'magma ocean'. While it was in liquid form, the lighter, lower density materials in the magma, mainly aluminium and silicates, floated to the surface and, as things cooled down, the whole outer crust solidified. A while later, around 4 billion years ago, the solar system entered a period called the 'Late Heavy Bombardment', when the Moon and other solar system bodies were peppered with asteroids and meteors. This bombardment caused the intense cratering all over the Moon's surface today.

The highlands are what are left of the original crust after this heavy bombardment. These areas are highly cratered and can be dated back to a time close to the Moon's formation. The highlands are pale in colour as they are made of the lighter materials that floated up through the magma. These regions are older than the Moon's other distinctive terrain, the maria.

The Maria

The maria (singular: 'mare') are the Moon's lunar planes, dark and relatively featureless compared with the highland regions. There were originally called maria – Latin for 'seas' – because early astronomers, looking at them from Earth, thought they were full of water.

During the period of heavy bombardment, the Moon, with no significant atmospheric protection, was hit hard. Immense asteroid impacts fractured the previously formed crust, allowing lava from the layers below to erupt to the surface through the deep cracks. This activity left huge pools of basalt lava on the lunar surface, which later solidified to form the maria. Basalt is one of the most common volcanic (igneous) rocks found here on Earth. It can be found on our ocean floors and around volcanic activity, such as on the islands of Hawaii. It is dark in colour to look at, hence the dark colouring of the maria, and is made up of about 50 per cent silica (silicon oxide). It may seem strange that a volcanic rock that is abundant here on Earth can also be found on the Moon nearly 400,000km away, but this gives us a clue to their probable common ancestry (see pages 47–9).

Radioactive dating has estimated the age of the maria at about 3 to 3.5 billion years old, so these plains are younger than the highland areas. The maria cover just over 15 per cent of the Moon's surface. Over the years, 23 maria have been identified and named. Most of these sit on the 'near side' of the Moon (the side that constantly points towards Earth, see page 53).

The variation in the Moon's surface, which I mentioned earlier, is not uniform. The highlands show evidence of older craters that have remained undisturbed, apart from the impingement of other craters over their surfaces. The maria, due to their more recent volcanic activity, tend to have less topographical variation.

Other Lunar Features

Rilles

The word 'rille' stems from the German for 'groove' or 'furrow'. Rilles were first identified and named by the German scientist Johann Hieronymus Schröter at the end of the eighteenth century. Rilles are cracks that can be seen on the lunar surface and they are thought to be produced by past volcanic activity on the Moon. They can often be tracked back along the Moon's surface to old volcanic vents and they may have been caused by collapsed lava tubes.

Domes

These are rounded, circular features that have gentle slopes and rise to an elevation of a few hundred metres. They are thought to have been formed by the flow of relatively thick lava erupting from vents. Being thick, the lava would have solidified before it travelled far, and formed the dome as it hardened. Domes typically have a diameter of about 10km, but can be as big as 20km across.

Wrinkle ridges

These are features found within the maria, caused by tectonic activity. When released, the basalt lava cooled and contracted, and in some places it did this at different rates, for example if there was a mix of different types of lava erupting from the same hole. This different rate of cooling caused a buckling of the surface and the formation of long ridges.

Grabens

'Graben' stems from the German word for 'trench' or 'ditch', and these too were formed from tectonic activity. They are essentially troughs, which are created when two cracks or faults that are roughly parallel to each other are stretched. The area that sits between the cracks subsides to form the graben. Most grabens are found within lunar maria near the edges of large impact basins.

Regolith: The Moon's 'Soil'

The surface of the Moon has had a tough time, having been pummelled by asteroids and meteors over much of its lifetime. This continuous pounding has broken down the top layers of its crusted surface, creating a lunar covering called regolith.

The term 'regolith' is used to describe a layer of unconsolidated material that sits on top of the bedrock of a planet or body. In the case of the Earth, the regolith is comprised of soil (organic plant remains in which new plants can grow), rock fragments, sand, volcanic ash and glacial drifts (material transported by the activities of glaciers). Yet the Moon's regolith has an entirely different composition, mainly made up of the Moon's own crust that was broken up by the asteroid and meteor bombardment.

Although some lunar regolith can be the size of boulders, 90 per cent of it has a grain size of less than 1mm. This means that most of the lunar surface material has a consistency that is finer than granulated sugar. The full mix would actually be closer to a combination of granulated, caster and icing sugar – but the chemical composition of the lunar regolith means that it would not taste as good!

Lunar soil is mainly made up of oxygen, with silicon, iron, calcium, aluminium and magnesium making up the bulk of the rest. It also has a sprinkle of more exotic elements like titanium, thorium and manganese. The proportions of these chemicals relative to each other are fairly similar to the proportions found on Earth. There is, however, a difference in chemical concentrations occurring in the lunar highlands and maria (see pages 26–8).

WHAT LIES BENEATH

Like the Earth, the interior of the Moon can be simply broken down into three main zones: the crust, the mantle and the core. Our knowledge of the interior structure of the Moon is still limited but the knowledge we do have has been obtained from missions that have been sent to the Moon over the last 50 years. The Apollo missions used a device called the Apollo Passive Seismic Experiment, made up of four seismometers, which were used to measure earthquakes and volcanic eruptions, and were deployed on the Moon's surface between 1969 and 1972. These seismometers took continuous measurements till 1977, generating a wealth of data on moonquakes (see pages 33–5) and other activities. Monitoring how seismic waves, generated by lunar activity, travel through the body has enabled scientists to calculate the probable structure of the interior of the Moon.

If we were to use a common everyday item to describe the Earth's internal structure, the item that would work best would be an egg. The eggshell represents the Earth's crust, a very thin layer covering the surface. The egg white represents the mantle, a deep volume that surrounds the egg yolk, which in this analogy would represent the Earth's core. This analogy works as the proportions of the egg layers approximately match the proportions of the Earth's interior zones.

However, the egg analogy does not work well for the Moon's structure. Here, a better analogy would be a rather boring sugar-coated chocolate-chip muffin. I say boring because the muffin has just a single chocolate chip at its centre, definitely a disappointing result for chocolate lovers everywhere. But when it comes to the Moon, it makes for an apt illustration. In this scenario, the sugar coating represents the crust, which in the Moon's case is the regolith that was mentioned earlier. Under the thin crust would sit the mantle, the fluffy cakey bit, and at roughly the centre would sit a relatively small core represented by the chocolate chip.

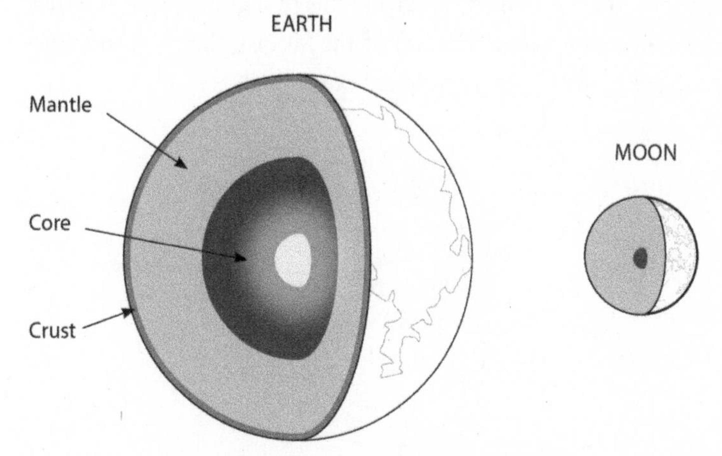

The different proportions of the inner layers of the Earth and Moon.

What Is the Moon Made of? (Clue: Not Cheese!)

We know a little about the chemical composition of the Moon, but the information we have has come from just two main sources. Firstly, the US Apollo and Soviet Luna missions, which returned a number of Moon samples from a small range of locations over the Moon's surface. In addition to this, scientists can use remote sensing, which is analysis of data gathered from orbiting around the Moon. The combination of these two methods has given us a general picture of the key elements that sit on the Moon's surface, as listed in the section above. However, neither of these methods is very helpful for seeing what's beneath the surface, so the elements that make up the interior of the Moon are not as well known.

It is thought that the centre of the Moon has a solid iron core, surrounded by a possibly molten, liquid outer core. This scenario is very similar to the Earth's structure. Next to this is a region that is partially liquid and it is thought that this is where moonquakes occur. The rest of the Moon – the crust and most of the mantle – forms a ridged outer shell.

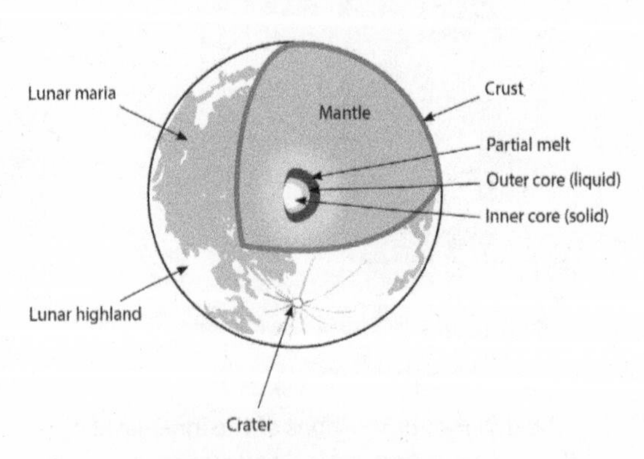

The internal structure of the Moon.

More of the Moon was in a molten state when it was first formed. During this period there was volcanic activity on the Moon, but this stopped as the Moon cooled and became more solidified.

Analysis of areas such as the maria, where objects from space (such as meteors and asteroids) have broken through the Moon's crust and released some of the material below, gives us an indication of the chemical composition of the lunar interior. The mantle is thought to be made up of two main types of basalt: olivine, which is a magnesium iron-rich silicate; and pyroxene, which contains silicon, aluminium and oxygen generally combined with calcium, iron or sodium. Some lunar basalts contain relatively high levels of titanium. Out of interest, olivine gets its name from the colour of its crystals. They are, as you might expect, a translucent yellowy-green olive in colour. This is the material that is used to make a gemstone called peridot. It is not one of the more common gemstones but I have a pair of earrings, which I am very fond of, that feature this mineral. The Moon might look dull and grey from a distance, but close up maybe there is the occasional flash of green.

MOONQUAKES

We generally think of the Moon as a stable rock, but surprisingly the Moon experiences quakes just like the Earth. In fact, recent re-analysis of the data obtained from the Apollo seismometers has revealed that the Moon is seismically active on an unanticipated scale.

There are thought to be four kinds of moonquake:

1) Deep moonquakes that occur 700km below the Moon's surface. These are caused by the gravitational bond between the Moon and Earth – see page 109.

2) Meteorite impact vibrations that are caused when the Moon is hit by meteors and asteroids.

3) Thermal quakes that are caused by the expansion and contraction of the lunar surface from the very severe temperature differences experienced over the Moon's surface – for example, at first light, when sunshine hits the cold lunar surface for the first time after two weeks of frigid darkness (see page 46).

4) The fourth sort of quake is a bit of a mystery. These are shallow moonquakes that occur just 20 to 30km below the Moon's surface. What is particularly interesting about these shallow moonquakes is their intensity: they can register as much as 5.5 on the Richter scale. The equivalent here on Earth would cause heavy furniture to move around and plaster to crack. Quakes on Earth usually die down after around half a minute as the waves get weakened by the presence of water in the rocks – the water, which is more compressible than the rock, acts as a sponge and absorbs the energy. The Moon, on the other hand, is quite dry, so when it experiences a shallow moonquake, it rings like a bell for some time – up to ten minutes on occasion.

The main mystery, though, is what is causing these moonquakes. If seismometers are placed carefully, it is possible to work out the source of a quake. However, the Apollo seismometers were placed at the various landing sites all on one side of the Moon and as such do not have the resolution to pinpoint the source of these shallow waves. Future missions are already being planned to deploy more seismometers at strategic locations across the Moon's surface, to get a better understanding of their origin. It is important to understand these phenomena as they could be significant for our future Moon bases and colonies.

AN ALIEN ENVIRONMENT

So what would it be like to live on the Moon? In this section we look at the challenging environment that the Moon would present to any future colonists (you never know, it could be you!).

The main factors that affect the Moon's environment are temperature, atmospheric pressure, sunlight and the solar wind, but we will also explore gravity on the lunar surface, the effect of its magnetic field and any seasonal changes one might experience. Welcome to the lunar landscape!

Temperature

From all the footage that we have seen from the Moon's surface you might be expecting a cold, dry environment. If so, then the diagram below may be somewhat surprising. It shows the mean temperature and temperature range of a) the Earth as a whole, b) Antarctica and c) the Moon. Here you can see that the mean temperature on the Moon is higher than that of Antarctica. We know that Antarctica is cold, but surely not lunar cold? To explain this unexpected result we need to look at what is happening at various locations on the Moon.

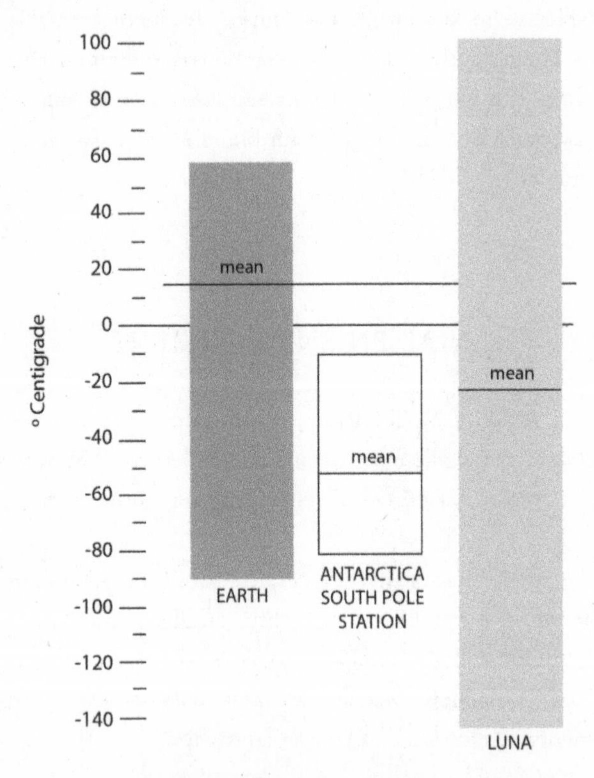

Planetary surface temperature

The lunar environment can be split into two very different zones: the polar and the non-polar regions. They can both experience enormous temperature differences, but this is particularly so in the non-polar areas. For example, at noon on the equator the temperature can soar to a scorching 100°C, but at night the temperature can plunge to –150°C. As the Moon has a very tenuous atmosphere these temperatures are localised, as there is no means of transporting heat to other areas apart from via radiation.

Meanwhile, at the poles the temperature variation is not so great, as these areas only receive sunlight at grazing angles, never direct. In the sunlit areas of the poles the temperatures may get as warm as -50°C, whereas in the areas that the sun does not reach at all, temperatures can be as low as -200°C and will stay at these levels forever. These sunless cold spots can act as traps for volatile chemicals. These chemicals may arrive at the Moon's surface in various ways but if they somehow end up in one of these cold spots then they have no means of escape, as they never heat up. They are trapped there forever. These areas are known as cold traps and make the Moon's poles very interesting places to study, as scientists sometimes discover chemicals that you would not expect to find there. It is thought that these cold traps may be one explanation for why water has recently been detected on the lunar surface.

Gravitational Field

We have all seen the footage of people walking on the Moon. You can see how strange their movements are as they attempt to travel around without tripping and bouncing across the lunar surface. Why is this? Well, simply put, the Moon has much less mass than the Earth and hence has a much smaller gravitational field. This means the astronauts' gait is changed considerably due to the lower gravitational force pulling on them.

To get a feel of how different it would be, take a look at the footage of Alan Shepard, from the Apollo 14 mission, playing golf on the Moon. It is an odd thing to watch – a man in a full space suit hitting golf balls across the lunar surface – and the action was criticised by some as a publicity stunt and a waste of time and money. However, it does show how the lower gravity affects things.

Shepard hit three balls using what I believe was a six-iron, but, due to the bulk of the suit, he swung with just one hand. As he watched the ball fly from his last shot he commented: 'Miles and miles and miles.'

But did it really travel that far? Well, on Earth a good swing would get a golf ball to a speed of around 180 miles per hour (290km/h) and this would be the same on the Moon. But on Earth such a shot would take the ball to around the 300 to 400-yard mark (about 320 metres: the world record for the longest golf drive is 515 yards or 471 metres). On the Moon there is virtually no atmosphere (so no air resistance) and the gravitational pull on the Moon's surface is just under 17 per cent of what we experience here on Earth. Combining both of these factors means that a swing on the Moon could carry a ball about 2.5 miles, or 4km. So Shepard's comment was quite accurate. As well as the amazing distance travelled, it would take around a minute for the ball to land. Life on the Moon for future colonists really will be a whole new ball game (excuse the pun).

But that is not the only surprising thing about the Moon's gravitational field. In an interesting discovery – which was fortunately made before the first Moon landings were attempted – scientists found that the Moon's gravitational field is lumpy. The lumpiness is caused by concentrations of mass that sit below the lunar surface, causing them to have a large positive gravitational anomaly. For the Moon landing this was critical information, as a change in the gravitational field could send a lunar probe wildly off course or cause it to plummet onto the Moon's surface. These geological structures have been given the name mascons, a shortening of the term 'mass concentrations', which sums them up nicely.

Recent surveys of the Moon have tried to work out the mystery of what lies below the surface causing these mascons. The current theory is that the bombardment history of the Moon is again the culprit. When extremely large asteroids hit the lunar surface, the impact was so hard that the resulting shockwaves penetrated into the Moon well beyond the crust. It seems that with some of these impacts, compression of the material around the impact zone caused these areas to increase in density, creating the mascons. Further work is needed to understand this, but for now we just need to be aware of these gravitational pitfalls for future missions to the Moon.

Magnetic Field

When the scientists started to analyse the Moon rock returned by the Apollo mission they were surprised to discover that some of the rocks were magnetised. At this stage, they had no idea that the Moon even had a magnetic field and were hard pushed to explain why a body of its size would have ever have generated such significant magnetism.

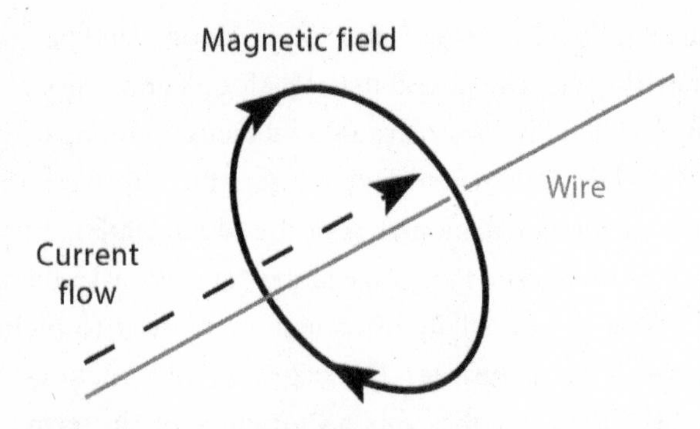

An electric current passing through a wire generates a magnetic field around the wire.

To get to the bottom of this, we first need to understand how the Earth's magnetic fields are generated. In a body or planet if certain conditions are met a magnetic field can be generated. If you conduct a simple experiment and pass an electric current through a wire, then a magnetic field is generated around the wire. (The opposite is also true, incidentally. If a wire is passed through a magnetic field then a current can be generated in the wire.)

The Earth has a solid metal iron and nickel core surrounded by a region called the outer core, which is made up of liquid metal, also made of iron and nickel. Near the centre of the Earth, both the inner and outer core experience temperatures so high that the metal should be molten. However, at the Earth's inner core, the very high pressures that exist there make it solid. As we leave the inner core of the Earth, the pressure drops enough to make the outer core liquid.

The metal inner core does not generate a magnetic field because it is solid, so there is no movement of the metal to generate electric currents. But the moving liquid outer core generates electric currents, as do differences in the temperature and pressure conditions experienced in the liquid outer core. These cause convection currents, just as we see with water and air in our day-to-day lives: even when we put on a pan of water to boil an egg, the water starts to move as it heats up. When a liquid or gas is heated, it decreases in density and rises up into cooler temperature zones; it then condenses and increases in density, sinking back into the hotter zones, only to be heated up again and continue the cycle. In addition, the outer core also moves due to forces generated by the Earth's daily rotation (known as the Coriolis force), which cause whirlpools within the liquid metal.

All these different movements cause currents to be generated, which in turn create a magnetic field. This process of a magnetic field being generated by a liquid or partially liquid core is known as 'geomagnetism'. Although the magnetic fields are being generated by different mechanisms, they all reinforce each other, creating an invisible but strong barrier that protects the Earth from the ravages of the solar wind (see pages 43–44).

The Moon, however, is much smaller than the Earth, so it was initially thought that if it ever had any magnetic field this would have been very short-lived. This seems to make sense, as a smaller body would have less pressure and lower temperatures at the centre – and also, being smaller, it would lose the heat that remained from its formation quite quickly. This would cause the core to solidify quickly and the magnetic field generation properties of a liquid metal core would have been lost.

However, recent analysis of an Apollo Moon rock has caused us to think again. The original magnetic Moon rocks that were examined were about 3.2 billion years old. This means that they were around in the first few billion years of the Moon's existence, when a molten layer was likely to exist. But the latest sample to be measured was just 2.5 billion years old – comparatively recent in geological terms. By this period it would have seemed likely that the liquid layer of the Moon would have solidified, so we would not expect a magnetic field to have existed then. However, measurement of these newer rocks revealed that they too were created in a magnetic period on the Moon. This means that the Moon's magnetic field must have persisted for around 2 billion years, a billion years longer than expected!

But what could have caused this extension to the magnetic life of the Moon? One of the strongest theories to explain this curious anomaly has its roots in the close relationship between the Earth and the Moon. The gravitational pull of the Earth on the Moon could have caused the Moon's core to heat, so the liquid metal stayed liquid for longer. It's an interesting idea but further investigation is needed to prove it.

The geomagnetic field generated by the Moon's liquid outer metal core finally did diminish and disappear altogether around 1.5 to 2 billion years ago, and as a result it was thought that there would be no measurable magnetic field on the lunar surface. However, recent magnetic measurements have now confirmed that the Moon *does* still have a very small field. Rather than being generated by geomagnetism though, it is thought to be caused by residual magnetism found in rocks on the Moon's surface. This form of magnetism is called crustal magnetism and it can also be found in areas here on Earth.

So why does any of this matter? To me it's fascinating how something as seemingly obscure as the magnetic properties of a handful of rocks can give us an insight into the origins of the celestial bodies, and how they were formed billions of years before our time. The Moon still has many secrets to unlock, of course, but bit by bit scientists are piecing together the jigsaw of how it came into being.

Atmosphere

It is often thought that the Moon has no atmosphere. This is understandable as it is so thin that it barely qualifies as an atmosphere, therefore is called an exosphere. If you took 1 cubic centimetre of the Earth's atmosphere at sea level, it would take a while but you could count around 10^{19} (that is 10 billion, billion) molecules contained within the volume. If the same thing were done at the surface of the Moon, 1 cubic centimetre would contain just 10^{6} molecules, just 1 million – a difference of 13 orders of magnitude. What's more, during the lunar night it gets so cold that these few molecules sink to the ground.

The chemical composition of the Moon's atmosphere is currently thought to contain the elements helium, argon, sodium and potassium. The helium is likely to have originated from the Sun and is transported to the lunar surface via the solar wind. This solar wind is a continuous stream of energetic, charged particles that flow outward from the Sun, mainly made up of protons and electrons. They travel at phenomenal speeds (up to 900km/s) and at temperatures of a whopping 1 million °C. They are caused by the huge temperatures at the outer layers of the Sun's atmosphere, an area called the 'corona'. These particles stream out deep into space, into a region called the heliosphere, which encompasses the solar system and extends out beyond the dwarf planet Pluto.

The solar wind causes some interesting phenomena. It can be seen during a total eclipse of the Sun as a halo around the obscured star. The solar wind also causes the northern and southern aurora on Earth, as these charged particles from the solar wind undergo a complex interaction with the Earth's magnetic field. And it causes the tails of comets to point away from the Sun, as this wind of charged particle 'blows' the tails in this direction.

Here on Earth, the magnetic field and a thick atmosphere limit our interaction with the solar wind. But on the Moon these lines of protection no longer exist, so the solar wind can deposit elements such as helium directly onto the lunar surface. Interaction with these highly energised particles and photons (packets of light) from the Sun also causes molecules to be knocked off the lunar surface to form its thin atmosphere.

Other sources for the tenuous lunar atmosphere include chemical reactions between the solar wind and material on the Moon's surface that has been liberated by impacts with meteors and asteroids and the release of gases from the lunar interior. Scientists are still puzzling over which of these various mechanisms contributes most to the lunar exosphere and further research is needed to find more answers.

How Long Is a Day on the Moon?

It seems like a straightforward question. On Earth it's easy: it's the time the Earth takes to do a complete rotation on its axis, i.e. 24 hours, isn't it? And during that period (if we ignore the

Earth's tilt for now), about half of that time will be spent with our hemisphere of the planet looking towards the Sun (hence daytime) and for about the other half of that period we will be looking away from the Sun, in our night time. But the Moon does not orbit the Sun, it orbits the Earth and it has a very strange relationship with us (see 'The Dark Side of the Moon', page 52). And the rotation of the Earth on its axis actually takes 23 hours, 56 minutes and 4.0916 seconds, if we are going to be precise. So what's going on?

Before we answer that, there is another question that needs addressing, which may sound even weirder – and that is 'what is a day?' In astronomy we talk about two kinds of day, the sidereal day and the synodic or solar day. But what is the difference?

The word 'sidereal' means in relationship to distant stars, so it is a measurement relative to the stars or constellations, rather than a measurement relative to local bodies such as the Sun, Moon or planets. To measure a sidereal day we would need to measure the time it takes for the stars to move through the night sky and return to their original position the following day. If we make this measurement, we can see the Earth actually takes the 23 hours, 56 minutes and 4.0916 seconds that I mentioned earlier. So what about our familiar 24 hours? Well, this figure is associated with a solar or synodic day. This is a measurement of the time it takes for the Sun to move through our skies and return to approximately the same spot the next day.

But why is measuring relative to the stars different from measuring relative to the Sun? When we measure a day relative to the Sun, because we are orbiting the Sun as well as rotating on our axis, it takes a few extra minutes every day for the Sun to sit in the same place in the sky as the day before: each day, we have travelled a little further around our solar orbit, so we need to turn an extra little bit to see the Sun in the same place. This extra time turns out to be around four extra minutes a day. But if we measure relative to the distant stars, we do not need to account for the travel around our orbit, so it takes less time for the stars to reappear in their original place.

So why have two systems? If we were to use sidereal time as our everyday clock then sunrises and sunsets would slip by about four minutes every day, which would be a bit of a pain to say the least. But if we used the solar day for astronomy then the position of the stars would slip over time, making tracking them a little more challenging; hence it's best to keep the two separate.

All this seems like a long way away from the Moon's daylight hours, you might think.

Well, in a similar way it takes the Moon 27 days, 7 hours, 43 minutes and 11.5 seconds (27.3 days) to orbit the Earth, and as the Moon is tidally locked to the Earth it also takes the same 27.3 days to rotate on it axis. But as the Earth is rotating around the Sun and again travelling along its orbit with the Moon in tow, it actually takes a little extra time for the Sun to reach the same position in the sky if you were looking at it from the Moon (as the Moon will also have travelled slightly further around its solar orbit in that time). Therefore the synodic or solar day on the Moon, the time between sunrise and another sunrise, is slightly longer than the 27.3 days – it actually takes 29.5 days.

So, on the Moon a day lasts about one Earth month, and it is daylight for just over two weeks followed by about two weeks of darkness. Another thing that future colonists will have to contend with!

Are There Seasons on the Moon?

There is a reason the Moon's climate seems so unchanging and bleak. We have the beauty of the changing seasons in some locations here on Earth because the Earth is tilted on its axis by 23.5 degrees. The result of our planet's tilt is that either the northern hemisphere or the southern hemisphere sits closer to the Sun as the Earth makes its year-long orbit around the Sun. The closer side receives more energy and this causes the change of season to the warmth of spring and summer. The Moon, by contrast, has an axial tilt of just 1.5 degrees and orbits the Earth rather than the Sun, so it experiences no significant seasonal changes.

HOW WAS THE MOON FORMED?

One of the many interesting things about the Moon is how it came about. We can see other moons but they're not like ours; for one thing our moon is much bigger in relation to the Earth than those belonging to other planets. So where did our moon come from?

Before the Apollo missions, there were three main theories as to how the Moon was created:

Moon Capture

The first was the idea that the Moon was a large, passing asteroid captured by the Earth's gravitational field. This seems to be a common occurrence in our solar system and explains the two moons of Mars. However, simple comparisons of the size of Phobos and Deimos compared to our companion demonstrated the flaw in this theory; the Moon is too large to fall into this category.

Co-formation

The next theory was that the Moon and Earth were both formed as part of the same event in the solar system. There are two possible ways this could have occurred. Like all the planets in the solar system, the Moon and Earth could have formed out of what's known as 'an accretion disc' – a spinning disc of materials containing gas, plasma, dust and particles that orbits around a central body, in this case the Sun. Over time, the material in the disc separated into bands and slowly clumped together to form two separate bodies, the Earth and the Moon. Due to the close proximity of the two, the Moon was then captured by the Earth's gravitational field and the Moon went into its orbit about the Earth. The other idea is that the Moon and Earth may have formed as a binary system in a localised accretion disc, with the two bodies spinning around each other from the get-go. Both of these theories would explain why the two bodies have similar compositions, as they were formed from similar areas in the accretion disk. But both theories have flaws too, the main one being the size of the Moon's core. If it were formed in this way, the Moon's core size should be a much bigger proportion of its overall volume, like that of the Earth's core.

Seeded from our Earth

The third theory is based on the idea that the early Earth was spinning so fast that a large blob of matter spun off it into orbit to form the Moon. However, the mechanics of this theory are hard to reproduce. If the Earth was spinning fast enough for a large blob to escape then it would seem more likely that the blob would have had enough momentum to carry it on out into space.

Collision Theory

Once the Apollo missions returned to Earth with its rock samples, the Moon formation theory went in a different direction. Chemical analysis of the samples showed a remarkable similarity between the Earth's composition and that of the Moon. It suggested some common heritage between them and the collision theory was born.

The collision theory suggests that a small planet about the size of Mars, hypothetically called Theia, collided with the early Earth in a glancing blow. The impact sent a ring of molten material into orbit about the Earth, which, over time, coalesced to form the Moon. It formed at just the right distance to be an independent body; any closer and Earth's gravity would have pulled the material back to Earth. It is an elegant theory and would explain the similarities between the chemical compositions of the Earth and Moon.

But there is a problem: the composition of the Earth and the Moon look too similar. If this collision occurred as predicted, the Moon should have more of Theia's material in its composition and should therefore show more variance than Earth. To date, no material that could have originated from Theia has been found.

Could the small variation in the composition of the samples be due to the limited range of areas that the samples were obtained from? The Apollo samples that were returned to Earth were, after all, obtained from a relatively small number of locations. Well, it would seem not because they are not the only samples of lunar material that we have obtained here on Earth.

Sites of Apollo and Luna landings on the Moon.

The Russian Luna programme returned a total 0.33kg of Moon samples to Earth and, as can be seen from the map, the combined Luna and Apollo sites are scattered quite widely over the Moon's near-side surface. In addition, samples from spacecraft are not our only source of lunar material. It can literally fall out of the sky. This is material that gets ejected off the Moon through asteroid and meteor impacts. If the impact is energetic enough then the ejecta can travel into space, eventually falling to Earth as a meteorite. The great thing about this process is that the material released this way comes from all over the Moon's surface, giving us samples from a wide range of locations. However, analysis of this material reveals a similar problem; the chemical composition of material obtained from the Moon is just too similar to the Earth's composition to explain the glancing-blow scenario with an entity like Theia.

So where does this leave the collision theory? Well, it still has a lot of support among scientists and it is still an active area of research, but many more lunar samples from known but widely varying locations would be a great help. I would like to offer my services to go and retrieve them, but, unfortunately, so far no one's taken me up on my offer!

Yet we do know some things for certain about the early days of the Moon. After it was first formed, the Moon was much closer to the Earth and days here were much shorter, taking no more than about five hours. As the Moon has moved further and further away from us, the Earth's rotation has slowed down. We've measured that the Moon is travelling away from us; it is moving at about the same rate our fingernails grow. In astronomical terms, that's actually quite fast! We know this because the Apollo astronauts left retroreflectors on the Moon's surface, angled mirrors, which means we can send laser light up to the Moon that gets reflected back. Timing that journey of light to and from the Moon means we can actually work out how the Moon is moving (see page 124). In fact, given a number of years (around a billion), it will be far enough away one day that the Earth could become unstable – and as the Earth's rotation slows down, interesting things could happen. For example, it seems likely the whole planet would wobble and gyrate!

We think this might have happened to Mars in the past and that it could be the future for the Earth. The poles might suddenly swing round to the equator, swapping places with it and causing catastrophic changes. The weather would be extreme, with fiercely hot summers and freezing, bitterly cold winters. How it will affect our planet depends on how quickly it happens; life can adapt in certain ways, but to adapt quickly and to such an extreme would be hard (see pages 133–4 for more on this).

Our planet's dependency on the Moon comes about because it's in a binary system; here in one hand is the Earth, here in the other is the Moon. The technical term for it is 'the conservation of angular momentum'. The Earth and Moon are both spinning together and, in a binary system, it's like two ice skaters going round and round; if they pull in towards each other, they go faster, whereas if they stretch out they start going more slowly. You can try this for yourself in a wheelie chair. Sit on it and spin yourself, holding a heavy weight. If you push the weight out, away from you, you'll slow down; whereas if you pull it in, you'll spin faster. Please be warned, I have seen one of my lecturers fly off a chair doing this demo so do take care. A helmet is advisable or, even better, watch the demo online: less fun but much safer.

THE DARK SIDE OF THE MOON

In astronomy there are many misnomers. One of my favourites is the term 'shooting star'. These can light up the night sky during any season but are more common during periods of meteor showers, when as many as 10 or 20 an hour can be seen. The name implies that one of the stars that we see in the night sky breaks free from its usual position and streaks across the sky.

In reality, the Sun, our local star, is over 1.3 billion times the volume of Earth – so stars don't tend to gad about much. Despite the name, stars have nothing to do with shooting stars. What we are actually seeing are small amounts of dust and debris burning up in the Earth's atmosphere. Meteor showers occur when the Earth orbits through a trail of dust and particles left behind as a comet passes through the inner solar system. As we travel by, many particles burn up in our atmosphere, causing the said shooting stars.

Another of my favourite misnomers is the term 'the dark side of the Moon'. It sounds quite menacing and conjures up all sorts of ideas. Why can't we see it and why is it dark? Well, just as with the shooting star there is no actual dark side of the Moon – but there is a whole hemisphere that we cannot view from our position here on Earth. It is more fittingly called the 'far side of the Moon' and this is all due to a phenomenon called tidal locking.

Tidal locking is quite common within our solar system; all the major moons in our solar system are tidally locked to their planets. A fine example of it is between the dwarf planet Pluto and it largest moon, Charon. In this case, Charon's orbit about Pluto takes the same amount of time as it takes the moon to rotate once on its axis. However, Pluto is also tidally locked to Charon so it always presents the same face to its moon too. It is like the two are staring at each other, never breaking their gaze, as they waltz around the solar system (creepy).

Tidal locking occurs due to the gravitational pull between a planet and its moon. As the Moon orbits the Earth, the Earth's strong gravitational force pulls at the Moon, causing it to elongate towards the Earth. When the Moon was first formed it was not tidally locked to Earth, and as it orbited the Earth a different face would point towards our planet. In those days we could see the full surface of the Moon over the course of time. But as each area got in close proximity to Earth, the gravitational force exerted by the Earth would deform the Moon, elongating the sphere and creating a bulge pointing towards the Earth's centre. Because the bulges took time to form, by the time they were in place the Moon's rotation would have moved them out of alignment, so the bulges were always slightly out of sync with the direction of pull towards the centre of the Earth. The fact that they were out of sync caused the bulges to act as gravitational handles or mass concentrations, which the Earth pulled on to bring back into alignment, the additional tugging causing the Moon's rotation to slow down. Eventually, when the Moon's rotational period had slowed down to matched its orbital period around the Earth, then the bulge was always in the same place, pointing directly to the centre of the Earth. At this stage, as the handles were now aligned with the pull, no further change in the rotational speed occurred and the Moon was tidally locked to the Earth, always presenting the same face.

The better named far side of the Moon was not seen in any detail until 1959 when the Soviet spacecraft Luna 3 flew past it and transmitted photographs back.

Interestingly, although the Moon is now tidally locked to the Earth, this process has not finished yet, as the Moon is slowly tidally locking the Earth to itself. This process is very, very slow and it is thought that the Earth won't be fully locked to the Moon for another 50 billion years, so don't hold your breath. When this does occur the Moon would not be seen to rise and set from Earth as it does now. Instead it would stay fixed in one position in the sky and only be visible from one hemisphere of the Earth. But considering that the Sun is only likely to be around in its current form for about 5 billion years, the tidal locking of the Earth is unlikely to happen before the end of the life of our solar system.

LIBRATION: SEEING WHAT LIES BEHIND

Libration is a phenomenon where even though the Moon is tidally locked to Earth, and only presents one hemisphere to an observer on Earth, it is possible to see more than 50 per cent of the Moon's surface from Earth. This happens because the Moon's orbit is not circular but elliptical, which causes a slow rocking back and forth of the Moon as it is observed from the Earth, allowing us to see slightly more of its surface at different times. It is also due to the misalignment of the Moon's rotational axis, which does not match up to that of the Earth.

Libration takes various forms: north to south, east to west and a diurnal or daily libration due to the rotation of the Earth over a day. Let's start with the east-to-west one first. This perceived movement happens because, although the Moon's rotation rate is constant, its orbital revolution rate around the Earth is not. If the Moon's orbit around the Earth were circular then its speed would be constant and the Moon would always present the same hemisphere to the Earth with no perceived libration in this direction. However, because the Moon's orbit is elliptical, when it is closer to the Earth it travels faster than when it is further away. Over an entire transit these fluctuations in the orbital speed average out, but at various points around the orbit the rotational speed can be relatively faster or slower than the spin speed.

When the Moon is closer to the Earth its orbital rate speeds up and is faster than its rotational rate. When this happens, the Moon rotates a little to the left, relative to the Earth, and a few additional degrees of the usually obscured right-hand side of the Moon can be observed. When the Moon is furthest from the Earth then its orbital speed is relatively slower than its rotational speed; this turns the Moon right, and reveals some of the usually obscured left side of the Moon. This form of libration gives us an additional 7-degree view of the Moon.

The north-to-south libration is smaller than the east-to-west one. It is caused by a misalignment between the Moon's axis of rotation relative to the plane of the Moon's orbit. This means that at different points in the Moon's orbit more of the lunar north and south poles can be seen as it travels slightly above us and slightly below us. This form of libration gains us just over an additional 6-degree view of the Moon.

The third type of libration is visible due to the rotation of the Earth. This movement means that in the course of a day we can get two slightly different viewpoints of the Moon around either side of the imaginary straight line joining the centres of the Earth and Moon. This gives an extra 1-degree view of the Moon.

All of these extra degrees or slithers of view caused by the librations add together to give lunar observers on Earth a 59 per cent view of the Moon, rather than the expected 50 per cent. In the past, before we were able to send probes into space, librations were critical in gleaning a little more information about the far side of the Moon. And even now they are of almost magical benefit to the amateur astronomer, allowing us to see what lies behind.

So now we've looked at the basic features of the Moon from a scientific point of view, let us look at it from another angle – a more personal one, perhaps. In the next chapter, I want to explore humankind's relationship to the Moon from an artistic and cultural perspective, and to delve into history to see how far our connection reaches back.

MOON PAST:

THE MOON IN OUR CULTURE

THE MOON plays such a prominent role in the night sky that people have observed it and considered it for many millennia. I am deeply fascinated by how the different cultures of the world have viewed the heavens, how the Moon has captured their imaginations and how it has influenced their lives since the beginning of history. In this section, I intend to go off-piste a little and indulge, if you don't mind, my enduring interest in a subject called archaeoastronomy: the investigation of the astronomical knowledge of prehistoric cultures. I also want to share some of the stories of the amazing people who have been inspired by the Moon through the ages, and the artefacts and places that give us insight into their knowledge. It is not a thorough, full or detailed look at the subject, but it is one that focuses on things that have caught my eye and remain close to my heart.

Let's start by considering a brief overview of humankind's long relationship with the Moon.

A BRIEF HISTORY OF MOONGAZING

Before we could travel to the Moon we had to look at it from afar and interpret what we saw. The timeline opposite shows what the people of the past learned about the Moon just by observing it.

Over time we have learned a great deal about the Moon and our understanding continues to evolve to this day. But we have remained fascinated and inspired by it too, as is evident in the way it has influenced our culture, art and lives. Let us look now in more detail at some of the people, places, artefacts and works of art that have been influenced by the Moon and which, in turn, have galvanised my own lifelong fascination with it.

~30,000 BCE France	The Abri Blanchard bone: this bone from the Aurignacian culture in Europe has markings thought to be associated with a lunar calendar. Though this is yet to be proved, if we are correct, this would be the earliest known example of its type (see pages 79–80).
~20,000 BCE Uganda/Democratic Republic of Congo	The Ishango bone: as well as what is arguably a tally of prime numbers, there are notches on this baboon bone that could mark the days of a lunar calendar (see page 80).
~8,000 BCE Scotland	Pits found in Aberdeenshire have been seen to recreate Moon phases in shadows, showing that nomadic hunter-gatherers were interested in lunar cycles (see page 74).
5th century BCE Babylon	Babylonian astronomers record the 18-year cycle known as the 'Saros period' that is used to predict eclipses, showing an understanding of the lunar cycles and possibly of solar–lunar alignment.
5th century BCE India	Indian astronomers record monthly lunar elongation. (The lunar elongation is the angle between the Sun and Moon with the Earth at the apex, i.e. during a full moon the lunar elongation is 180 degrees and during a new moon the elongation is 0 degrees.)

4th century BCE **Shi Shen** **China**	Astronomer Shi Shen gives lectures on the prediction of solar and lunar eclipses.
428 BCE **Anaxagoras** **Greece**	Astronomer Anaxagoras suggests that the Sun and Moon are both big balls of rock and that we see the Moon due to its reflecting sunlight.
400 BCE **China**	Chinese astronomers also become aware that the Moon is seen due to light from the Sun reflected from its surface.
310 – 230 BCE **Aristarchus** **Greece**	First recorded calculation of Earth–Moon distance, with Aristarchus obtaining a value of about 20 times the radius of Earth.
***c.* 150 BCE** **Seleucus of Seleucia** **Mesopotamia**	Seleucus correctly theorises that tides are due to the attraction of the Moon, and that their height depends on the Moon's position relative to the Sun.
~50 BCE **Jing Fang** **China**	Musical theorist and astronomer Jing Fang notes that the Moon is spherical, as opposed to being a disc as some believed.
***c.* 100–168 CE** **Ptolemy** **Greece**	The Earth–Moon distance figures are greatly improved by Ptolemy's calculations of a mean distance of 59 times the Earth's radius and a Moon diameter of 0.292 Earth diameters, which are close to the correct values (60 and 0.273 respectively).

499 CE **Aryabhata** **India**	The mathematician-astronomer Aryabhata mentions in his *Aryabhatiya* that reflected sunlight is the cause of the shining of the Moon.
***c.* 950–1009** **Ibn Yunus** **Egypt**	Ibn Yunus, an Egyptian astronomer, uses stars observed during a lunar eclipse to work out the longitude of a position on the Earth's surface (see page 69).
1031–1095 **Shen Kuo** **China**	Shen Kuo of the Song dynasty creates an allegory to explain the waxing and waning phases of the Moon. According to Shen, it is comparable to a round ball of reflective silver which, when doused with white powder and viewed from the side, appears to be a crescent.
1610 **Galileo Galilei** **Italy**	Using a telescope of his own design, Galileo draws one of the first telescopic sketches of the Moon, which he includes in his book *Sidereus Nuncius* ('Starry Messenger'). From his observations, he notes that the Moon is not smooth, but has mountains and craters (see page 70).
1651 **Giambattista Riccioli and Francesco Grimaldi** **Italy**	Riccioli and Grimaldi devise their own Latin naming system for many features on the Moon, published in their book *Almagestum Novum*. These names are still in use today (see page 143).

1753 **Roger Boscovich** **Croatia**	Boscovich, a Croatian astronomer and mathematician, argues that the Moon has a negligible atmosphere, as stars that are occulted by the Moon (i.e. they are seen to travel behind it) disappear instantly rather than fading.
1757 **Alexis Clairaut**	The French astronomer Clairaut is able to calculate the first accurate mass of the Moon.
1834 and 1837 **Wilhelm Beer and Johann Heinrich von Mädler** **Germany**	The map and book created by these German astronomers are the first accurate trigonometric studies of lunar features, and include the heights of more than 1,000 lunar mountains.
1870s **Richard Proctor** **England**	Proctor proposes that the Moon's craters were formed by collisions, as opposed to the prevalent theory that they had been formed by volcanic activity (see page 25). This view gains support throughout the remainder of the nineteenth century and, in the early twentieth century, leads to the development of lunar stratigraphy – part of the growing field of astrogeology.

FIVE PEOPLE

Looking back through time, there have been some amazing people who have gazed up at the Moon. Some have done it in the spirit of scientific enquiry, whereas for others it has been purely for pleasure. In this section I want to consider five people who have mused over our local orb. Of course, people have been doing this for millennia but as we travel further and further back over the years, very few records are kept of moongazers. Some cultures with mainly oral traditions, slip through the net, as the records of their observations are lost to the annals of time. Bearing this in mind, I give you my five people, filtered through the sieve of history.

En-hedu-ana: Astronomer Priestess of the Moon Goddess (*c.* 2354 BCE)

The translation of this priestess's name means 'ornament of the heavens'. Her father was the Sargon, the progenitor of the Sargonic dynasty in Ur (then in Mesopotamia, now Iraq) around 4,000 years ago. And hers is the first female name recorded in history and the first poet known by name too.

En-hedu-ana was appointed the chief astronomer priestess of the Moon goddess of the city, which meant that she managed the city's great temple complex as well as the extensive agricultural precinct around it. Her work entailed numerous astronomical duties, including tracking the position and the phases of the Moon. I have always thought that her job and title sounded really cool, so if any city is looking for an astronomer priestess of the Moon goddess, please let me know and I will send through a CV.

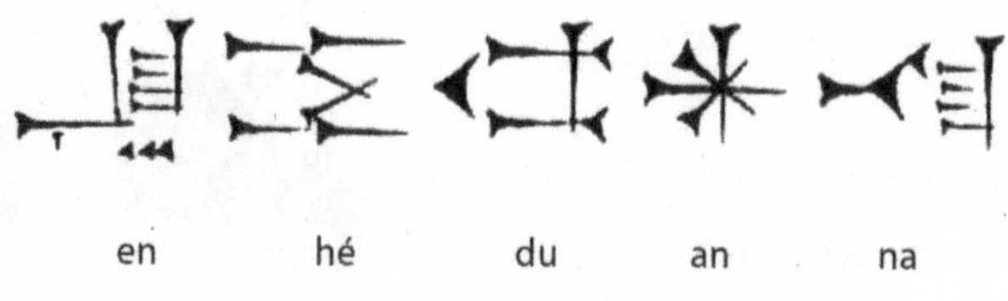

Sumerian transcription of En-hedu-ana's name.

There are numerous images of her, but at least one of the busts labelled with her name has a substantial beard. It was thought that this was a status symbol, which she wore when addressing her fellows. (I like to point out that it may have taken 4,000 years, but I am lucky that I do not have to don such accoutrements when participating in *The Sky at Night*.)

As well as having her astronomical career, she was also a prolific poet who wrote many hymns, which were used in Mesopotamia for centuries after her death. She also wrote some prayers; these inspired some to call her the Shakespeare of her time. Just over 40 of her poems survive today. In these poems, she often wrote about her duties as astronomer priestess, including using the heavens to make decisions on Earth. In one such poem she also described a real woman as possessing 'exceeding wisdom'. An amazing woman and a true role model for all of us.

Thales Predicts a Solar Eclipse (585 BCE)

It sounds like the plot of some blockbuster, epic movie. It was 28 May 585 BCE. It started like any other day. Alyattes, King of Lydia, was battling against Cyaxares, King of Media, near the River Halys (now

 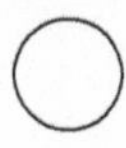

central Turkey). This ding-dong war had been going on for some 15 years; something had to be done to bring the fighting to an end.

A single man steps out into the fray – but he is not just any man; he is an astronomer, and only he has the key to breaking the deadlock. Equipped only with his knowledge of the heavens, he makes a prediction that will stop the warriors in their tracks and resolve the war once and for all.

OK, that was the hammed-up, Hollywood version of events, but many are not sure if the Greek version written by the historian Herodotus is technically correct either. He wrote that Thales of Miletus, an astronomer of the time, predicted an eclipse during a year in which the Medes and the Lydians were at war:

> [A]nd as they still carried on the war with equally balanced fortune, in the sixth year a battle took place in which it happened, when the fight had begun, that suddenly the day became night. And this change of the day Thales the Milesian had foretold to the Ionians laying down as a limit this very year in which the change took place. The Lydians however and the Medes, when they saw that it had become night instead of day, ceased from their fighting and were much more eager both of them that peace should be made between them.

It's not much to go on but with our understanding of how the solar system works we can dial the clockwork universe backwards and see if and when this was likely to have occurred. When looking at solar eclipse occurrences around the time that Thales lived, the most likely candidate is 28 May, 585 BCE. Some believe it might have been an eclipse 25 years earlier in 610 BCE but the jury is still out on this one, and with no more information available I guess we will never really know.

However, predicting a solar eclipse is not quite as easy as I have made it sound. In a solar eclipse, the Moon's shadow across the Earth is narrow, and the maximum duration of totality at any given place is just over seven minutes.

Today we can do these calculations with computers, but the early Greeks did not have all the information needed to make a prediction like this with any accuracy, and apart from Thales' forecast, there are no other records of the Greeks of this era accurately predicting eclipses.

One explanation is that Thales is believed to have studied the Egyptians' techniques of land measurement and astronomy. Could he have used their knowledge to make the prediction or did he simply find out about the eclipse from the Egyptians?

However he came about his knowledge, his prediction of the eclipsed Sun had the desired effect, as it was widely interpreted as an omen. When it came to pass, with day turning suddenly to night, the two warring factions agreed on a peace deal and used the River Halys as the border between their two kingdoms. A happy Hollywood ending for the story.

Ibn Yunus: Ahead of his Time (*c.* 950–1009)

Ibn Yunus was an Egyptian, Muslim, astronomer and mathematician. He is mainly remembered as the author of the 'Hakemite Tables', a detailed catalogue of astronomical events that covered a period of around 200 years, including his own observations of two solar and one lunar eclipse in 977, 978 and 979. He worked under the patronage of the Fatimid dynasty for 26 years at the time of the founding of the city of Cairo.

He is less well known for perfecting a method of determining the longitude of a location by observing stars during a lunar eclipse. This method was not used in Europe till much later.

Although overlooked by many scholars in the history of astronomy, his work has now been recognised and a crater on the Moon has been named after him.

Galileo Galilei Builds a Telescope (1564–1642)

Galileo is commonly believed to have been the first person to have viewed the Moon through a telescope, but an English nobleman may have beaten him to it ... (see page 82).

Our story begins with a young Dutch spectacle-maker called Hans Lippershey, who in 1608 applied for a patent for his spyglass design. One anecdote goes that Lippershey's children were playing with lenses in his shop when one noticed that a distant weather vane looked much closer when viewed through two of the lenses. The patent was rejected as it was thought that the design was already in common use, but nonetheless by some means a description of Lippershey's spyglass design was brought to the attention of Galileo Galilei, who was a professor of mathematics at the University of Padua.

Galileo took the design and by 1609 he had made his first telescope with a magnifying power of around nine. Although already an amazing improvement on the naked eye, by December 1609 Galileo had come up with an improved design, which he pointed at the Moon. Now he was able to observe it as never before and noted its rough surface with craters and mountains. By 1610 he wrote up his observation of the Moon and other views of the solar system in a short book, *Sidereus Nuncius*, which also included 70 sketches of his observations.

The book was met with mixed reviews; the public seemed charmed by what they read, but fellow astronomers were in uproar. Galileo's rough Moon surface did not comply with the doctrine taught by the Church. Heavenly bodies, made by God, were by their very nature perfect, smooth, spherical bodies. This view was not just held by the Church but had been perceived as the way of things since Aristotle some 2,000 years before.

This discrepancy between observation and what was considered to be true perhaps marks the start of Galileo's troubles with the Church, but by building his telescope he was able to make a great leap forward in our understanding of the solar system for ever.

GALILEO AND THE CHURCH

The fracture between Galileo and the Catholic Church was to become a source of embarrassment for the Vatican in years to come. In 1632, Galileo published a book called the *Dialogue Concerning the Two Chief World Systems*, in the form of a discussion among three people: one who supports Copernicus's heliocentric (Sun-centred) theory of the universe, one who argues against it, and one who is impartial. Though Galileo claimed his *Dialogue* was neutral, it was clearly not. The advocate of Aristotelian belief – who defends the Church's position – comes across as a simpleton, getting caught in his own arguments.

Unsurprisingly, the Church took great umbrage at the book, and Galileo was called to face the Inquisition and put on trial. In 1633, he was found guilty of 'vehement suspicion of heresy', put under house arrest (which he remained under for the rest of his life), and his *Dialogue* and all of his other works were banned, including any as-yet unwritten.

Just a few years later the Church accepted the Copernican theory of the universe but maintained the ban on Galileo's book until 1820, and it was not until 31 October 1992 – 350 years after his death – that Galileo received an official apology from Pope John Paul II. The apology was for putting him on trial in the first place.

Wáng Zhēnyí Listens to the Moon (1768–97)

Wáng Zhēnyí was a scientist from the Qing dynasty era who wrote a number of articles on astronomical phenomena and conducted practical hands-on experiments to understand what was going on in the solar system. One of her areas of interest was studying the phenomenon of the lunar eclipse. For this, as well as using historical records, she measured and calculated lunar positions and set up experiments to study the movements of the celestial bodies. To do this, she placed a round table in a garden pavilion to act as the Earth; she hung a crystal lamp on a cord from the ceiling beams, which represented the Sun; then on one side of the table she had a round mirror as the Moon. She could now move these three objects around as if they were the heavenly bodies to see how they interacted with each other. After much studying of astronomical principles, she illustrated that a lunar eclipse happens when the Moon moves into the Earth's shadow. She wrote a paper on her findings entitled 'The Explanation of a Lunar Eclipse', and her observations were later found to be very accurate.

Although many of her findings had been discovered much earlier and other Chinese astronomers had previously tried to show that eclipses were natural phenomena, it was still common practice to attribute them to angry gods, so it was an extraordinary achievement for a self-educated woman to attain that level of knowledge in this environment. But she was not short of self-belief – she was a strong supporter of equal opportunity for both men and women, as a line from one of her poems testifies, 'daughters can also be heroic'.

Wáng Zhēnyí brought her poetic sensibility to her astronomy as well. In her 'Journal of Listening to the Moon Pavilion', she wrote: 'Some would say, "The Moon cannot be heard." Alas, honestly, it cannot be heard. Yet something at its centre may enlighten its listeners. Therefore, the Moon can be listened to.'

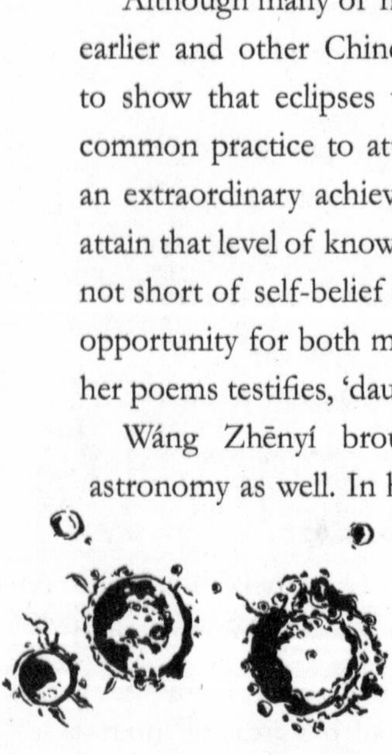

Zhēnyí shows all the markings of a fellow lunatic and, I think, is to be admired for it. Unfortunately, her life was short, as she lived only to the age of 29, but in that time she was able to make a real contribution and change perceptions of women in the feudal society in which she lived.

FIVE PLACES

The Sun has inspired the construction of numerous monuments around the world. Many of them are familiar to us today, such as Stonehenge, Newgrange or the Ra temple discovered in a Cairo suburb in 2006. In a similar way, the Moon has led to the creation of some impressive ancient tributes too. Several of the surviving ones are associated with the measurement of time passing.

Measuring the passing of time is a useful exercise and one that many early cultures were interested in. Clues from naturally occurring events gave an indication of the regular rhythms of nature and were noticed by early humans. The change in the weather linked with the seasons, the flooding of rivers, the growing cycles of plants, the migration of birds and animals and the movement of the stars … all of these things indicated the passing of time.

One of the clearest indications of time passing, which many early humans latched on to, was the waxing and waning of the Moon. Many of the earliest lunar monuments recorded the change of Moon phase, providing a sort of public seasonal calendar with a timescale that could be easily interpreted by those versed in astronomical knowledge. Here is a brief introduction to some of them, which I hope whets your appetite to find out more.

Warren Field in Aberdeenshire, Scotland (~8,000 BCE)

In 2004, a number of pits were noticed in an Aberdeenshire field when an aerial survey of the area picked up some unusual crop markings. After extensive excavation, a series of 12 significant pits were discovered, which stretched from the field across a hilly terrain covering a length of over 50km.

The structures are thought to mimic the phases of the Moon so as to track the lunar month over the course of a year. Effectively they serve as a giant time-keeping device.

In fact, this device turned out to be even more sophisticated than first realised, when archaeologists noticed that the whole structure aligned with the mid-winter solstice. It is thought that it was built this way to allow for corrections between the lunar year and the solar year. Astonishingly, the extremely early date of these pits – they are believed to be 10,000 years old – means that they were built by hunter-gatherers, rather than the more settled farmers that we usually associate with monument building. It means that this humble Scottish field is host to one of the oldest lunar calendars found anywhere in the world, predating the Mesopotamian time-reckoners by nearly 5,000 years.

The Knowth Complex in County Meath, Ireland (~3,200 BCE)

Newgrange is a prehistoric site of great astronomical interest situated in Ireland. The 85m mound is a passage tomb encircled by kerbstones, each decorated with a spiral design. During the winter solstice, light streams into the main chamber through a carefully aligned roof box near the entrance to the complex. It is famous the world over and attracts many tourists. However,

Newgrange was built to celebrate the Sun; we are more interested in its less well-known sister site, Knowth, which was built in homage to the Moon.

Knowth is thought to contain one quarter of all the megalithic art found in Europe today. Its main mound is of similar size to Newgrange but it has two passages leading to the central body, which are believed to be aligned to lunar activities.

It seems that the builders of this complex had a good working knowledge of the movements of the Moon. This allowed them to predict eclipses and other lunar phenomena. Activity at the site is thought to date back some 6,000 years, with at least 12 phases of activity there spanning many years.

One of the most extraordinary features of Knowth is a carving of what is believed to be the oldest lunar map in the world, at some 4,800 years old. To give some context, the next oldest description of the lunar maria known to science is that created by Leonardo da Vinci, over three millennia later.

The carving in the eastern chamber at Knowth (shown right) is believed to be a lunar map. Transposed onto a naked-eye sketch of the Moon (shown left and centre) you can see that the marks align with the lunar maria.

The Callanish Stones in the Isle of Lewis, Scotland (~2,900 BCE)

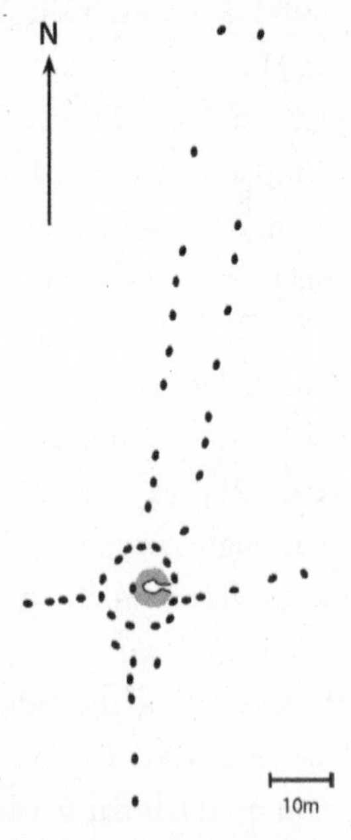

Diagram showing the placement of the Callanish Stones.

The stones of Callanish have stood at this site for around 5,000 years, predating Stonehenge by about 500 years. As well as a circle of upright stones in the centre, there is also a series of stones arranged around it in a cruciform shape, with an 'avenue' of parallel stones running to the north. It is believed that the ancient Britons erected them as an observatory for tracking a rare astronomical phenomenon called the 'lunar standstill', which occurs every 18.5 years.

If this theory is true, it was an extraordinary endeavour, requiring incredibly accurate observations of the Moon kept over decades and then a huge amount of effort to get the stones into the right position. And, of course, they would only be able to check their alignment once every 18.5 years. The Callanish Stones say a great deal about the role of the Moon in the lives of these ancient people. They clearly venerated the Moon and thought that its position in the sky was highly significant.

LUNAR STANDSTILL

Taking place in an 18.61-year cycle, this is a phenomenon that corresponds to the moment when the declination – that is, the angular distance of the Moon's orbit north or south of the celestial equator – of the Moon's orbital path (as compared to the plane of the Earth's equator) is at its maximum. When this happens, an observer on Earth will see the Moon's position in the sky changing over the space of two weeks, going from rising very high in the sky to skimming low over the horizon.

Ziggurat of Ur in Mesopotamia, now Iraq (~2,100 BCE)

The Sumerian name for this ziggurat means 'temple whose foundation creates aura'. The original temple complex featured a stepped pyramid (the ziggurat) and other monuments that were built to mimic the mountains where it was thought that the gods lived, creating a home away from home for them in the city of Ur. Some of the earliest structures built on this site were commissioned by King Ur-Nammu to honour the Moon god Nanna/Sin. By the sixth century BCE the complex was in ruins, but then the last king of the Neo-Babylonian empire, King Nabonidus, rebuilt the complex even larger than before. Interestingly, the very early foundations of the ziggurat are believed to be the complex that was presided over by En-hedu-ana, astronomer priestess of the Moon goddess mentioned earlier (see pages 65–67). It would seem that the Moon was worshipped on this spot for hundreds of years. The foundations of the complex still exist and can be visited today.

Temple of the Moon in Beijing, China (1530 CE)

This temple was used until 1911 by the emperors of the Ming and Qing dynasties, who would make animal sacrifices to the Moon god at this site at the autumn equinox. At other times it was used by army commanders and lesser nobles. Contained within the temple, sitting on a dais, is the Altar of the Moon, where the sacrifices were made. Within the complex are facilities to support a smallholding, with a storehouse, a kitchen, an area for washing the animals and a freshwater well. The complex even included a changing room and rest area for the Emperor to use during the period of divinations. The temple was opened to the public in 1955 and is surrounded by a park visited by locals and tourists alike.

FIVE ARTEFACTS

We can search the few ancient records that survive to find out about those who lived before. In some cases we have the opportunity to walk in the places where people of the past lived and worshipped. But my preferred form of archaeology is to look at the artefacts of the past. From these items we can gain knowledge of what our ancestors achieved and the level of skills they obtained.

Artefacts often give you a glimpse into a different type of history too. The people that we hear of through the span of history are usually the great and the good. The places that are preserved are usually the halls and grand palaces. But with artefacts we can sometimes gain a glimpse of the everyday past by looking at the sort of things that ordinary people like you and I might have owned.

Aurignacian Lunar Calendar, France (~30,000 BCE)

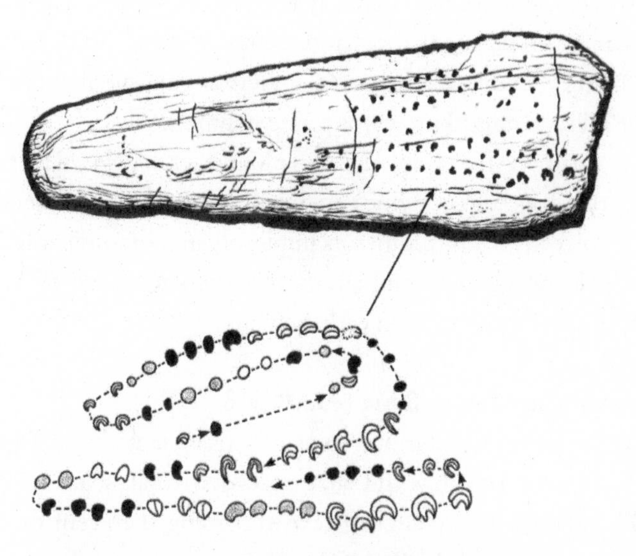

Sketch of the markings on the Abri Blanchard bone.

The Aurignacian bone found at Abri Blanchard in France is thought to have representations of the phases of the Moon carved into its surface, showing an interest in the lunar cycles dating back millennia. The Palaeolithic archaeologist Alexander Marshack was the first to realise that these marks are likely to be more than just ancient doodles. If his theory is correct, the Aurignacian bone is the earliest lunar calendar to be found anywhere in the world.

Ishango Bone, Uganda/Democratic Republic of Congo (~20,000 BCE)

In 1960, a baboon's fibula bone was found in the Belgian Congo (the site is now on the border of the Democratic Republic of Congo and Uganda). The bone is significant due to the distinctive tally marks made over its surface. There has been much debate as to what these marks mean. Some believe that they reveal a mathematical understanding and that one set of tallies seems to represent prime numbers.

Analysing the bone, Alexander Marshack suggested that the marks on its surface represented six months of a lunar calendar. If correct, this is another example of an extraordinarily old time-keeping device and shows that the knowledge of lunar cycles was truly global in early humans.

Ugarit Clay Tablet, Syria (~1,375 BCE)

A clay tablet found in the ruins of the ancient city of Ugarit documents possibly the first recorded total solar eclipse. The author of the ancient text, recording the event on the tablet, described the Sun as being 'put to shame' during the eclipse. It is thought that the eclipse occurred on 3 May

1,375 BCE and lasted for a duration of two minutes and seven seconds. However, more sophisticated dating of the tablet and mention within the text that Mars was visible during the eclipse now indicate that the more likely date was 5 March 1,223 BCE. Either way, it is one of the earliest solar eclipses ever recorded.

Antikythera Mechanism, Greece (~87 BCE)

In 1902, an interesting artefact was found in the wreck of a Roman ship off the coast of the Greek island of Antikythera. The object was intriguing, with its intricate mechanical clockwork mechanism made up of over 30 interacting bronze gear wheels. But at the time no one knew what it was. The device had dimensions of about 30cm by 20cm by 10cm and was housed in a wooden box. Initially it was all fused together as one incomplete lump, but over time it was separated out in to three distinct bodies.

Not knowing exactly what it was, for many years it was described as the first analogue computer. But with further investigation using X-ray, gamma-ray and CT scans, a replica of the mechanism was made in 2008. It turned out to be an orrery (a clockwork model of the solar system), including the Sun, the Moon and the five planets that were known at the time. It is thought to have been used to predict astronomical events like solar and lunar eclipses.

Further analysis of the original mechanism has revealed that inscriptions on the faceplate displayed both the Greek zodiac and the Egyptian calendar. One of the lower dials indicates the months in which lunar and solar eclipses were expected.

The knowledge of the technology behind this object was lost at some point over the following generations, and works of similar complexity did not appear again until the invention of the mechanical astronomical clocks of fourteenth-century Europe. The mechanism was truly well ahead of its time and there is still much speculation as to who could have made it.

The First Moon Drawing (But It Is Not by Galileo), Britain, 1609

For my next artefact we take a huge leap in time. In December 1609, Galileo made a telescope with a X20 magnification. With this he was able to view and draw the Moon with a level of detail that had never been seen before.

However, in England at around the same time, the British astronomer Thomas Harriot (1560–1621) was also using a telescope he had constructed to produce a magnifying power of six. With this telescope Thomas Harriot viewed the Moon and made a sketch dated 26 July 1609. Although Galileo seems to have made his first telescope around the same time, he did not produce a dated sketch of his observations until after the construction of his second telescope in December that year, over four months later.

So Harriot definitely created the first dated sketch of the Moon, but who was the first to look at the Moon through a telescope? Many claim it was Harriot but from the evidence I have seen so far, I think that it could go either way.

FIVE POEMS

Poetry almost by its very name invokes emotions in all of us. In a very few words, it has the ability to make us think and feel.

In this section I want to explore how poetry has depicted the Moon, and how that poetry makes us feel.

'Midnight Poem' by Sappho of Lesbos

My first poem is by Sappho of Lesbos; she is one of the few female poets whose writing survives from ancient times. We do not know much about her but believe that she lived from around 630–612 BCE to 570 BCE. She was very highly regarded in her time and what is left of her poetry is still studied today. Plato described her as the 'Tenth Muse', and her likeness has been depicted in busts and on pots. The reason I mention her here is because of a fragment of a poem that has been attributed to her. It is called the 'Midnight Poem', and reads:

> The moon has set
> And the Pleiades;
> It is midnight,
> The time is going by,
> And I sleep alone.

With beautiful economy, she sums up a restless night on her own, but even more remarkably, from the few astronomical details expressed in the poem, astronomers Professor Manfred Cuntz and UTA Planetarium director Levent Gurdemir were able to do some astronomical detective work and estimate what time of year the poem was likely to have been written.

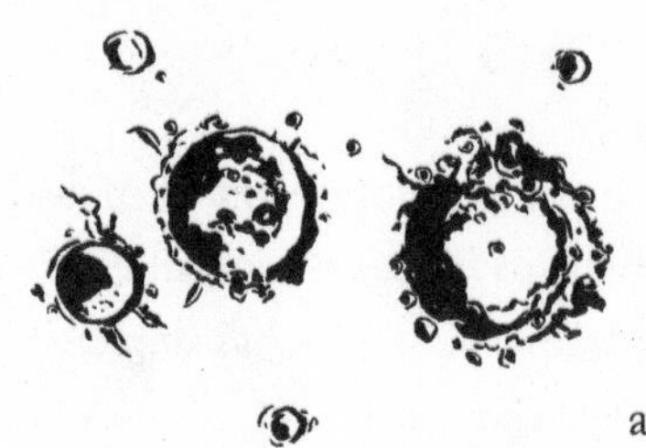

The two key pieces of information were the setting of the Pleiades – an open star cluster visible to the naked eye – at the time of midnight, and the likely location of where the poem was written (the Greek island of Lesbos, where Sappho reportedly spent most of her time). Calculations revealed that in 570 BCE, the Pleiades set below the horizon at midnight on 25 January, making it the first date the poem could have been written. Each night after this date the Pleiades set four minutes earlier. We can only ever get a ballpark figure for this, of course, as although the poem says 'midnight', the water-clock technology that was used at the time could be significantly out. Taking this into account, the latest date that the poem was likely to have been written was 6 April, as by this date the Pleiades set before the skies were dark enough to see them. Although an estimate of the day and month of the poem's penning can be attained, the exact year is harder to define as a similar situation would have occurred a few years earlier. However, in any chosen year there are only a few dates when the Moon would have set shortly before midnight.

It seems amazing that little is known about the poet's life, but we can get a reasonable guestimate as to when she wrote her poem from the mechanical, clockwork nature of the universe.

'Drinking Alone in the Moonlight' by Li Po

My next poem is by Li Po, thought to have lived between 701 and 762 CE. He lived in the Tang dynasty in what is often called the golden age of Chinese poetry. His works are as widely regarded today as they were in his own time.

'Drinking Alone in the Moonlight' is one of his best-known poems. The lines that caught my eye are written below:

> ... lifting my cup,
> I asked the Moon
> to drink with me ...

I have been in this situation many times before, for instance at the Gemini South Observatory in Cerro Pachón in Chile, working on the telescope during the day to install the bHROS instrument (a high-resolution optical spectograph). Later, I would be on my own at night with a glass of something enjoyably local, saluting the Moon for company. This poem captures that moment wonderfully and it feels as if we lunatics have been around and enjoying our moments of companionship with the Moon for many, many hundreds of years.

As an interesting footnote, legend has it that the poet Li Po died by drowning when he reached from his boat to grasp the Moon's reflection in the river. Whether it is true or not, there is no clearer illustration of the seductive power of the Moon's reflected light.

'A Hymn to the Moon' by Lady Mary Wortley Montagu

I feel that antiquity has not been particularly kind to women, so to include three in my lunar poetry tribute is a real treat. My third poet is Lady Mary Wortley Montagu (1689–1762) and her poem 'A Hymn to the Moon':

Thou silver deity of secret night,
Direct my footsteps through the woodland shade;
Thou conscious witness of unknown delight,
The Lover's guardian, and the Muse's aid!

By thy pale beams I solitary rove,
To thee my tender grief confide;
Serenely sweet you gild the silent grove,
My friend, my goddess, and my guide.

E'en thee, fair queen, from thy amazing height,
The charms of young Endymion drew;
Veil'd with the mantle of concealing night;
With all thy greatness and thy coldness too.

Montagu was a remarkable woman: as well as her writing, she is also celebrated for introducing and promoting the concept of smallpox inoculation in Britain after witnessing it in her travels in the Ottoman Empire, some 80 years before Edward Jenner developed his vaccine. She had limited success but did get some members of the royal family involved.

Her poem 'A Hymn to the Moon' is wonderful: short and succinct yet full of emotion. My favourite line, 'My friend, my goddess, and my guide', sums up for me the relationship many of us have with the Moon. We admire it from afar; we are mesmerised by its beauty. We share drinks with it when lonely, finding it a companion in even the remotest of places. And we use it as a guide. In my life the latter has been so true. The

 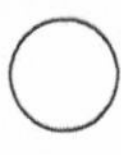

Moon helps us all to travel across the darkened Earth at night when we have left behind the streetlights' glare, and for me the Moon has influenced my whole life and set me on a somewhat unexpected career path.

'Child Moon' by Carl Sandburg

This poem by the Swedish-American poet Carl Sandburg (1878–1967) beautifully captures a child's sense of wonder when viewing the distant, silent, yellow orb of the Moon in the night sky, and watching its light filtering through the leaves. The end of the poem describes the child falling asleep babbling about what she has seen.

I mentioned earlier that much of my lunacy comes from my father's love of the orb, but 'Child Moon' matches so well my experiences with my own daughter. Following in the family tradition, she too is mesmerised by the Moon. Now we scan the skies together for that small glimpse and she has taken on a fascination of everything Moon-related, even insisting sometimes that I call her Luna.

To me it brings up the age-old question of nature or nurture. Could there be a genetic predisposition here for lunacy, passed on from father to daughter to daughter? Or is my daughter's interest purely a matter of indoctrination? But there is a third factor at play: the Moon has mass appeal for children, and perhaps that is where the true beauty of the Moon lies – the fact that it has the ability to bring out the child in all of us.

'The Woman in the Moon' by Carol Ann Duffy

The British poet Carol Ann Duffy put together a beautiful anthology of Moon poems past and present in 2009. She shows many signs of being a fellow lunatic and her anthology *To the Moon* has some amazing gems. One of the poems takes on the persona of the Woman in the Moon, sorrowfully looking back at Earth and seeing the destruction wrought by humankind on the forests and seas.

I love this poem and the fact that it brings out the female aspect of the Moon. It is something that I have always found very appealing: she is the ideal role model, powerful, serene and essentially in charge of her own destiny, very much like my mother who from an early age needed to make her own way through life and who showed me how to be a strong, independent woman.

The last phrase 'Darlings, what have you done … to the Earth?' sums up one of the most powerful legacies of getting into space and travelling to the Moon: the concept of the 'fragile Earth'. The famous photograph *Earthrise* epitomises this. It was taken by Bill Anders during the Apollo 8 mission (the first human mission to leave Earth's orbit and orbit the Moon). As he and his companions Frank Borman and James Lovell were travelling over the lunar surface they were able to see the Earth rising and to capture the beautiful image on film for posterity.

An audio recording of their conversation captured their excitement at the view:

Anders: Oh my God! Look at that picture over there! There's the Earth coming up. Wow, is that pretty.
Borman (joking): Hey, don't take that, it's not scheduled.
Anders (laughs): You got a colour film, Jim? Hand me that roll of colour quick, would you?
Lovell: Oh man, that's great!

As we run around on the Earth's surface, it is easy to forget that we are travelling on an orb in space. Yet early photographs taken of the Earth from the Moon, such as *Earthrise*, showed us how truly vulnerable we are. Many believe that images such as this were one of the main stimuli that created the environmental movement, making us a bit more aware of the impact we are having on our planet and causing us to ask ourselves, 'What have we done to the Earth?'

So the Moon, beautiful in itself, also affords us a degree of introspection as we gaze back at ourselves from its orbit.

Contemplating the Moon's beauty as revealed in photography leads us nicely into our next section, which shows how it has also been the inspiration for storytellers throughout the ages.

FIVE FOLKTALES AND SCIENCE-FICTION STORIES

Science fiction is one of my favourite genres. It was one of the things that got me reading when, as a dyslexic child, there seemed little point in making the effort to struggle through the pages. Science fiction told me such amazing stories that it was worth the endeavour.

There are also a couple of folktales in this section in order to include cultures with a more oral tradition of storytelling – and so many such legends are centred on the Moon and try to interpret its movement across the sky. In the same vein, many science-fiction stories feature the Moon: as our obviously closest celestial neighbour it seems like the ideal location for an adventure. In this section, therefore, I have included examples from both genres. All the stories still resonate with me today, even though one of the stories mentioned is over 2,000 years old.

Wan Hu (Date Unknown)

I feel that this ancient Chinese legend needs to be included here as it shows someone so caught up in their admiration of the Moon that they go to great lengths to get there.

The story goes that a Chinese official named Wan Hu was so mesmerised by the Moon that he decided to travel off the Earth to go and visit it. Using the technology available at the time, the bureaucrat commanded that some 47 fireworks (bamboo tubes filled with gunpowder) should be attached to his wicker chair to propel him lunarward. The underlings duly did as they were told. They attached the rockets, lit the touch paper and then retreated to a safe distance.

There was an almighty bang and thick clouds of smoke bellowed outward. When the smoke had finally cleared ... Wan Hu was nowhere to be seen. Had he made it?

It seems likely that this is a moral tale warning of desiring something out of reach. Some date the Wan Hu story as far back as 2,000 BCE, but this is nearly 2,500 years before gunpowder was invented so it is likely that the legend is more modern.

But to answer the question of whether Wan Hu did make it, the answer is surprisingly yes!

In 1965, a spacecraft called Zond 3, which was originally destined for Mars, was recommissioned as a lunar space probe. It was a success: it was able to take the first pictures of the 30 per cent of the far side of the Moon that up till then had not been seen. As a tribute to the legend that is Wan Hu, a crater on that far side, 52km wide and 5km deep, was named after him. The name was made official by the International Astronomical Union in 1970, so Wan Hu made it to the surface of the Moon after all.

Bahloo the Moon (Date Unknown)

This is an Australian aboriginal folktale explaining the difference between humans and the Moon. In aborigine legends, the Sun is female and the Moon is male.

Bahloo the Moon would sometimes step down out of the night sky and wander about the Earth when people were sleeping. He would take his pet snakes with him on these sojourns. One day he came to a river. Not wanting to get his snakes wet, he asked some locals to carry them across the river for him. Although the people liked Bahloo, they were afraid of his snakes so refused to help. Angered by this, Bahloo used some floating bark to cross the river, keeping his snakes dry. He then took a stone and threw this into the water, and said to the people: 'You are like this stone: when you die you will sink and never be seen again. Whereas I am like the bark; I will always rise again.' From this time on, people hated snakes as they were a reminder of Bahloo's words. But Bahloo kept on sending snakes to humans to remind them that they did not help him when he needed them.

In this tale we see an explanation in folklore for why the Moon rises again and again each night.

A *True Story* by Lucian of Samosata (*c.* 125–180 CE)

Lucian was a Syrian writer living under the Roman Empire who wrote satirical stories in Greek. His book begins by saying that, despite the title, its contents are 'notorious lies delivered persuasively'. It is thought that it was written in response to the far-fetched fantasies of authors such as Homer, and that Lucian wanted to show that believing in such outlandish tales was foolish. His story is credited by many to be the first example of a science-fiction tale, which is why I have included it here.

The plot follows a group of shipmates who get caught up in a whirlwind and are transported to the Moon's surface, where they find themselves in the middle of a war between the Moon King and the Sun King, both of whom want to take over Venus, the morning star. Lucian describes the aliens as being strange hybrid creatures. I won't say who wins the war but in typical mythical tradition, when the travellers return to Earth, their tribulations are far from over.

***Somnium* by Johannes Kepler (1571–1630)**

Kepler was a mathematician, astronomer and astrologer, and a key player in the scientific revolution of the seventeenth century. The cosmologist and science-fiction writer Carl Sagan described him as 'the first astrophysicist and the last scientific astrologer', which sums up Kepler's position, living as he did just on the cusp of the age of scientific reasoning.

He is much better known for his scientific works, including his laws of planetary motion, but as well as his scientific endeavours, he was the author of a number of science-fiction stories.

In *Somnium*, meaning 'the dream', Kepler's mind wanders into a dream after reading about Libussa, a Czech princess who became queen of her people. As Kepler put it, 'When I came to the story of the heroine Libussa, so celebrated in the magical arts, something happened that night. After contemplating the stars and the moon, I settled into bed and fell into a deep sleep.'

In his slumber Kepler fuses fact (the astronomer Tycho Brahe features in the tale), fiction (his characters make a journey to the Moon) and fantasy (they are carried to the Moon by demons). The story relates the adventures of an Icelandic boy whose mother has the ability to use spirits/demons to transport them to the land of Levania, aka the Moon:

> The island of Levania is located fifty thousand German miles high up in the sky. The route to get to there from here, or back to this Earth, is rarely open. When it is open, it is easy for our kind, at least, to travel. But transporting humans is truly difficult, and risks the greatest dangers to life.

Kepler then goes on to describe the geography of the Moon and he is clearly aware of the challenges of living on the near side and far side of the Moon in relation to the Earth.

> Therefore, as geographers divide our sphere of the Earth into five zones according to their celestial phenomena so is Levania divided into two hemispheres: one of these is the Subvolvan, the other is the Privolvan. The Subvolvans are forever blessed by the light from Volva which for them takes the place of our Moon. But the Privolvans are eternally deprived of any sight of the Earth.

The story is an interesting mix of science fiction and fantasy that makes for a fascinating read. Although written in 1608, it was actually published posthumously by Kepler's son in 1634.

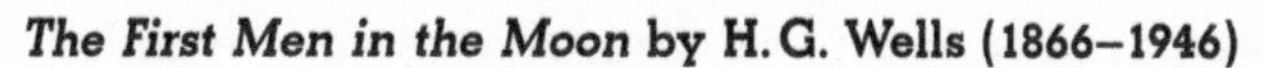

The First Men in the Moon by H.G. Wells (1866–1946)

In contrast to Kepler, Wells is a well-known science-fiction writer; his most famous book is probably *The War of the Worlds*, which describes an alien invasion by Martians. *The First Men in the Moon* is a bit of a departure from the usual Moon story as the aliens in this case live in the Moon's interior. This may have been due to the emergence of large telescopes, already established in Wells's lifetime. At the time of publication in 1901, the largest telescope in the world was the aptly named Leviathan of Parsonstown, a 72-inch reflecting telescope in Ireland (72 inches is the diameter of the main mirror of the telescope). With a telescope of this size, observations of the Moon would have revealed that there were no significant structures on the lunar surface, so having a civilisation inside the Moon would have made much more sense (or possibly on the far side of the Moon, not visible from Earth).

The novel centres on two adventurers named Bedford and Cavor, who decide to attempt an expedition to the Moon. The means by which the two heroes propel themselves into space is an anti-gravity material named 'cavorite'. Although this idea seemed novel for the time, Wells was actually accused of stealing it from other authors.

En route to the Moon, Wells describes weightlessness and on arrival on the surface he discusses the extreme temperatures that exist between the day and night sides of the Moon. He also imagines the aliens that live inside the Moon. The first one they encounter is a mooncalf, which he describes as follows:

> First of all impressions was its enormous size; the girth of its body was some fourscore feet, its length perhaps two hundred. Its sides rose and fell with its laboured breathing. I perceived that its gigantic, flabby body lay along the ground, and that its skin was of a corrugated white, dappling into blackness along the backbone. But of its feet we saw nothing. I think also that we saw then the profile at least of the almost brainless head, with its fat-encumbered neck, its slobbering omnivorous mouth, its little nostrils, and tight shut eyes. (For the mooncalf invariably shuts its eyes in the presence of the Sun.)

Later, they spot a Selenite, a more sophisticated type of alien:

> He presented himself ... as a compact, bristling creature, having much of the quality of a complicated insect, with whip-like tentacles and a clanging arm projecting from his shining cylindrical body case. The form of his head was hidden by his enormous many-spiked helmet – we discovered afterwards that he used the spikes for prodding refractory mooncalves – and a pair of goggles of darkened glass, set very much at the side, gave a bird-like quality to the metallic apparatus that covered his face. His arms did not project beyond his body case, and he carried himself upon short legs that, wrapped though they were in warm coverings, seemed to our terrestrial eyes inordinately flimsy.

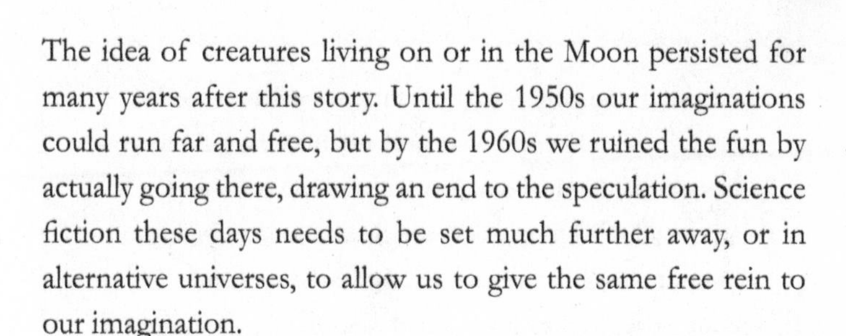

The idea of creatures living on or in the Moon persisted for many years after this story. Until the 1950s our imaginations could run far and free, but by the 1960s we ruined the fun by actually going there, drawing an end to the speculation. Science fiction these days needs to be set much further away, or in alternative universes, to allow us to give the same free rein to our imagination.

DOES THE MOON MAKE US MAD?

Lunacy

I mention a number of times in this book that I am a self-certified lunatic. My father would use this as a term of endearment and it has been used by my family ever since. To us, it means we are stimulated and inspired by the Moon. Yet there is a long history of the idea of the Moon influencing human behaviour and unleashing a darker side in us all too.

The word 'lunar' stems from the Latin word for moon – *luna* – and Luna was also the Roman goddess of the Moon. In turn, the term 'lunatic' derives from the Latin *lunaticus* meaning 'moonstruck' and was used to refer to a madness caused by the Moon. This belief was relatively popular and the philosophers Aristotle and Pliny the Elder both theorised that, as the Moon affects the tides, it could also influence the brain.

Even in the twenty-first century, we are bombarded with cultural references that indicate that the full moon affects human behaviour. But is there any evidence to support this? Well, in 2007, the Sussex police deployed more patrols on the streets of Brighton during full moons because they have noted a correlation between violent incidents and the phases of the Moon. In 2011, investigations conducted by the *World Journal of Surgery* also found that more than 40 per cent of medical staff believed lunar phases had an impact on human behaviour. Yet for every paper supporting this theory, there are several others that use the same research conditions and find no significant pattern.

Our belief in the connection between crime and the Moon could well stem from the past, when, before the advent of artificial light, it would have been easier both to commit crimes and to witness them under the brightness of the full moon. However, a phenomenon called 'illusory correlation', as coined by the University of Wisconsin, could also be responsible. This is the perception of an association that does not in fact exist. It sounds obvious, but as humans we are more likely to recall 'significant events' that happen rather than insignificant events. When the Moon is full and something noteworthy happens, we pay more attention than when nothing happens when the Moon is full. We then share these stories and this reinforces the idea of strange happenings at the time of the full moon.

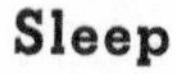

Sleep

Some research has indicated that our sleep patterns can be more restless during the time of a full moon, even when we are unaware of the phase the Moon is in. In 2013, Swiss researchers also found that, on average, people slept 20 minutes less during a full moon; 20 minutes may not seem like much but for some disorders (like bipolar) it can trigger an episode. Other researchers have conducted similar tests but found no correlation, so I guess that the jury is still out on this one!

Animals

Although the effects of the Moon on humans are not fully understood, the effects on animals are well documented. The lunar cycle and particularly the full moon are utilised by the animal kingdom to synchronise events and facilitate visual communication in nocturnal animals. Corals across the world synchronise the release of eggs and sperm into the world's oceans on a given night, which happens on or close to a full moon.

Other creatures behave differently around the time of full moon too. Lions are thought more likely to attack humans during the days around the full moon. This could be explained by the fact the lion's usual prey takes cover because of the additional light, while humans can get out more due to the better illumination.

Werewolves

The legend of the werewolf – the shape-shifting of a human into a wolf-like creature, usually when there is a full moon – has a place in folklore across the world. One of the first recorded mentions of a transformation of a person into a wolf is in the *Epic of Gilgamesh*, a text written around 2,000 BCE. In the story, a goddess turns a shepherd into a wolf as she grows bored with his attentions. This means that his dogs and sheep turn against him and he comes to a grisly end, torn apart by his dogs.

The legend of the werewolf is still alive and kicking today, and almost always associated with the Moon. One of my favourite depictions is in *An American Werewolf in London*. This 1981 film, written and directed by John Landis, is full of dark humour and violence. It terrified me when I was young but also enchanted me. One of the aspects of the film that I have since come to appreciate greatly is that the songs that make up its soundtrack all have a Moon connection, from Creedence Clearwater Revival's 'Bad Moon Rising' to Van Morrison's 'Moondance' (an old favourite of mine), as well as several versions of the popular classic, 'Blue Moon'.

FIVE WORKS OF ART

As I am a scientist, some may think it strange that I wish to wax lyrical about works of art, but this is based on a relatively recent concept that the arts and science don't mix much. This dichotomy was brilliantly discussed by the physicist Richard Feynman in 1981 when he was interviewed for a television programme shown on the BBC. In the interview he told of an ongoing disagreement he was having with a friend.

His friend, who was an artist, argued that, as a scientist, Feynman had little appreciation for the beauty of, for instance, a flower. The friend said that, as an artist, he himself could comprehend the flower's intricate beauty whereas, as a scientist, Feynman wanted only to dissect it.

Feynman admitted that his visual appreciation of a flower may be limited in comparison to a trained artist, but that he could still marvel at its beauty.

However, Feynman's marvelling did not stop at the flower's appearance. As a trained scientist, he could analyse the flower and infer that the flower had evolved particular visual markings in order to attract insects. Also, he knew the flower could convert carbon dioxide and water into sugars – a reaction that is the base of the food for all life here on Earth.

So, in many ways, Feynman's admiration of the flower went even further than that of his friend as, in addition to his awe at the flower's beauty, his scientific analysis delved further to appreciate the wonder of the flower's scientific make up. Therefore, rather than his scientific standpoint detracting from the beauty and awe of the visual, it enhanced it.

To my mind, I think that the argument goes both ways: a knowledge of science can augment the appreciation of the beauty of a thing and a knowledge of art can sometimes lead to a better understanding of the science we explore. The key is to make both accessible to all who are interested.

Art depicting the Moon, like the Moon itself, can fill me with an awe and wonder that I barely understand. In this section I would like to mention a few works of art which have had that impact or have proved to be of interest to my inner lunatic.

***The Crucifixion* and Other Paintings by Jan van Eyck (1390–1444)**

The prize for the first astronomically accurate representation of the Moon in a piece of art had been assigned for many years to Leonardo da Vinci, but recent analysis of the work of the Flemish artist Jan van Eyck has shown that he included accurate details of the Moon's maria and highland areas in a series of three paintings dated as early as 1420. The mystery is how van Eyck managed to obtain such a detailed knowledge of the Moon's surface. Did he have access to some form of magnification? Unfortunately, the mystery may remain, as none of his notebooks have survived.

Sketches of the Moon by Leonardo da Vinci (1452–1519)

Leonardo's sketches of the Moon caused him to investigate an interesting phenomenon called earthshine or 'ashen glow' (see page 147). It is a magical thing to observe: when you look up at a crescent moon you can still see the faint outline and glow of the rest of the moon beside it. This glow occurs when light from the Sun is reflected from the Earth, hits the Moon and is reflected back to Earth where we observe it.

This was witnessed by Leonardo and from his observations he was able to work out the principle of what was happening. He wrote up his findings in a volume later called the *Codex Leicester* (named after Thomas Coke, Earl of Leicester, who purchased it in 1719).

Leonardo went on to conclude that the Moon was covered in water, which he believed was the reason why it reflected the Sun's light. Although this idea was incorrect he did rightly suggest that the Moon has its own gravity, which kept the water in place rather than allowing it to fall to Earth. He describes how if the water were to fall to Earth, then so should the Moon. As he put it: 'Therefore not falling, it is clear proof that the water up there and the earth are supported with their other elements, just as the heavy and the light elements down here are supported in a space that is lighter than themselves.'

Keen observation and artistic flare enabled da Vinci to see the scientific principles behind everything he drew.

***An Experiment on a Bird in the Air Pump* by Joseph Wright of Derby (1734–1797)**

This famous picture was painted by Joseph Wright of Derby in 1768. It depicts a travelling scientist demonstrating how to make a vacuum by removing air from a flask containing a cockatoo, who will surely die if the experiment continues much longer. It may seem like a strange picture to include in a book on the Moon as the Moon is not a prominent feature of the painting. However, I love the work of this artist. Many of his pictures

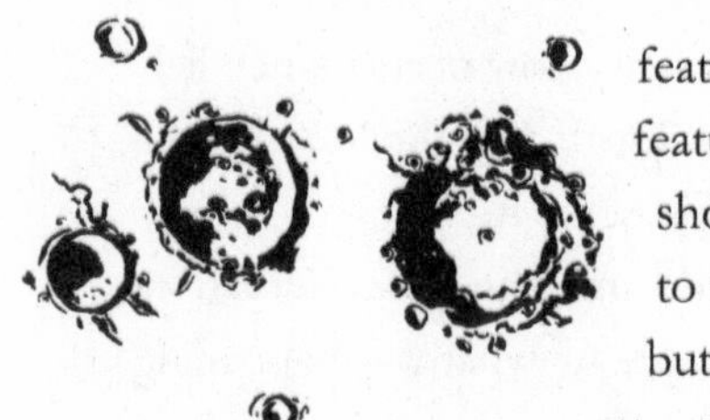

feature the Moon and many of them feature scientific experiments being shown to an audience. This appeals to the science communicator in me, but what has always grabbed my attention is his use of light. In many of his paintings we see a crowd of people lit by a single light-source near the centre of the image. This gives us a feeling of enclosure, of being part of the group.

He also manages to include lots of emotions in this particular image. The young woman hiding behind her hand, turning away from the experiment in horror or disgust. The girl looking up with interest but fear as she clings on to her neighbour. Two others in the picture appear to be having a quiet chat, and an older man looks on with the wisdom of knowing what happens next. It feels like a real moment in time frozen for our pleasure. And through the opened window, partly hidden in the cloud, a full moon looks down on the proceedings. To my eye, it is a wonderful marriage of science, art and emotion.

Moonscape by Roy Lichtenstein (1923–97)

Roy Lichtenstein was a famous American pop artist who was a contemporary of Andy Warhol and Jasper Johns. His most famous works include *Whaam!* and *Drowning Girl*. His work had a distinctive comic-book style that became celebrated across the world, so it may seem perhaps a strange choice of artist for something as ethereal as the Moon. But currently hanging in Tate Liverpool is a screenprint on plastic called *Moonscape*.

Lichtenstein had an ocean-fronted house in Long Island and it is thought that the views from this location inspired some of his dramatic seascapes. The thing I find fascinating about this picture is that, although the Moon is quite passive and the cloud is rendered very much in Lichtenstein's traditional cartoon-print style, the sea is a thing to behold, dynamic and always moving. In fact, viewing the image in books or online does not do it justice and to me each photograph of the picture looks unique. I include it here because it shows how a subject as traditional as the Moon can be interpreted in so many different ways. It is a very modern take on something we have stared at for millennia.

Over the Moon: An Optical Illusion by Rob Gonsalves (1959–2017)

Rob Gonsalves was a Canadian surrealist artist who described his work as magical realism. He was inspired by the works of fellow artists like Escher and Dalí, artists whom I also admire, and created beautiful worlds that I for one would love to jump into. Unsurprisingly, given the nature of his work, the Moon often featured in his paintings. The picture that caught my eye is titled *Over the Moon* and it depicts a girl on a swing in a darkening evening sky. Her legs are high in the air and her head and body are tilted back as far as they can go. She has a delighted smile on her face as she seemingly defies gravity.

Her swing is attached to a tree and behind it a full moon can be seen with lots of detail painted over the Moon's surface. It is haloed by clouds which, as we travel across the painting, turn into a huge magnified Moon with details of craters and maria shown on its surface. The background for this vast Moon is the dark view of space. The girl has swung so far she has got out there.

This image brings home many childhood memories for me, including the sheer joy of powering the swinging motion, when it truly felt for a few moments as if I was defying gravity. My ever-present thought was that, if I could swing hard enough, I could launch myself into space. This picture evokes the spirit of the Moon and the child in me, and makes me smile every time I see it.

Sadly, Rob Gonsalves took his own life in 2017, but he leaves a legacy that can inspire us all.

In this chapter we have been able to explore a very different relationship with our moon – not the clinical, scientific relationship governed by the rules of physics, but an emotional one, focusing on how the Moon makes us feel and act. It is a relationship that, as we can see here, extends back to virtually the start of the human race – but will it continue? As we live increasingly in a world with 24-hour light and with more and more distractions from the natural world, are we losing our long-standing connection with our celestial partner?

Well, if we look at our children, they still have a natural fascination with the night sky, as demonstrated in Sandburg's poem. Maybe it is up to adults to keep the lunar light burning within, enjoying its presence as we once did as children.

MOON PRESENT:

A SHARPER FOCUS

WHAT'S THE Moon ever done for us?

It's the classic Monty Python question; I mean what *has* it done for us? We all have a vague notion that it gives us the tides, but how else can a ball of rock in space help us here on Earth?

I've always loved the Moon, but I think that like many people I have taken it a little for granted, not appreciating all that it does for us. So in this chapter, I want to explore the relationship we have with our closest neighbour in more depth. I want to find out what the Moon does for us day to day, and what would happen here on Earth if the Moon were closer or farther away.

THE RAW POWER OF THE MOON

Let's start with the basics: the tides. We have all seen the effects of rising or falling water on the beach or in a river perhaps, and we can clearly observe high tide and low tide at differing times of day. But to see one of the great tidal wonders of the world we need to travel to the west coast of Scotland, to a place called the Falls of Lora.

Here, every six hours water from the Atlantic Ocean pours into Loch Etive and every six hours it pours out again. All of it – 45 million tonnes – has to pass through a narrow channel at the Connel Bridge, creating superfast currents and white-water rapids.

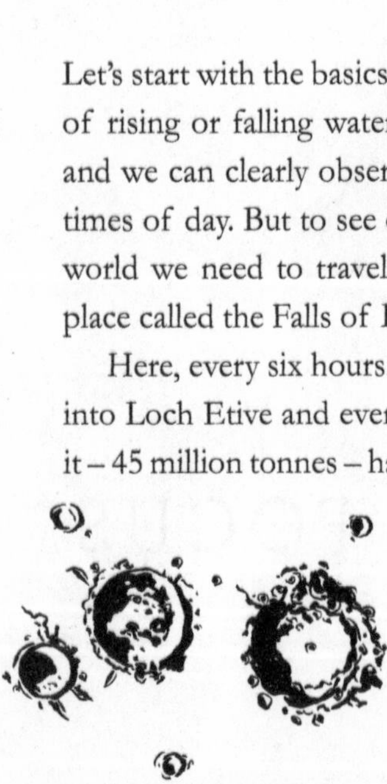

Expert kayakers come from all over the world to take advantage of this lunar phenomenon. They spin, jump and capsize their boats as they

 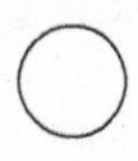

battle the falls. They have to fight hard not to be swept away. So what is happening out in space that can move around such a large amount of water down here on our planet?

Well, it is all about gravity. The Earth exerts a gravitational pull on the Moon, keeping it in orbit; but in turn, the Moon exerts a gravitational pull on the Earth, tugging at the world's oceans, creating what's known as a tidal bulge.

Although the tides are mainly caused by the Moon, the Sun's gravity plays a part in the motion of the tides too. Sometimes it works to enhance the influence of the Moon and sometimes it works to diminish it, but we will cover that later. Let's keep things simple for now and look at the Earth/Moon system.

The Moon has about a hundredth of the mass of the Earth but, because it is located relatively close to us, it exerts a significant force. As the Earth rotates, the 'pull' of the Moon affects the part of the Earth directly below it. When land is below, we don't notice a significant movement, but if it is an ocean below then the Moon's gravitational force can move the water, causing it to bulge towards the Moon.

The size of the bulge is not just dependent on the size of the gravitational force but also on the difference in the gravitational pull of one side of the planet compared with the other. On the side of the Earth near the Moon, the gravitational force (Fmoon near) is 7 per cent bigger than that experienced on the far side of the Earth (Fmoon far).

EARTH–MOON SYSTEM (NOT TO SCALE)

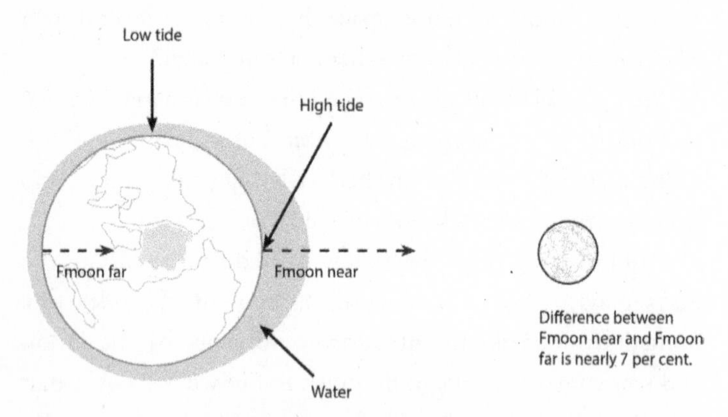

Diagram showing 'differential force', causing a single water bulge on Earth.

This is called the 'tidal force', but it is not just associated with tides. It is also responsible for a range of different phenomena, including spaghettification in black holes (the process by which objects get extremely elongated as they fall in). I prefer to call it the 'differential force' to keep it more general than just tides. As the Earth rotates and passes through the bulge, that location will experience high tides. At points perpendicular to this, away from the bulge (see diagram), low tide will be experienced as the water is dragged away to the high tide.

But the system is more complex than this. This scenario would give us just one high tide a day but most places experience two. Why? Well, the reason is that, in addition to the bulge of water in the direction of the Moon, there is an equivalent bulge on the opposite side of the planet. But what is causing this?

As well as the Moon's gravitational force, there is another force at play. This occurs due to the Earth and Moon's rotation. We think of the Moon orbiting the Earth and it does, but it does not orbit around the centre of the Earth. Instead, both the Earth and Moon orbit around a common point that lies between them. This point is called the 'barycentre' and is the centre of mass of the two bodies.

One of the things that keeps coming up when we talk about the Moon is that it is quite large compared with Earth. Because our moon is large in comparison with our planet, the point of common rotation does not sit at the Earth's centre. In fact, it's about 4,700km from the centre, but still well within the planet's radius of 6,400km.

Let's go back to the analogy of the two ice skaters spinning around each other that I mentioned in the first chapter (see page 52). If you have two equal-mass adult skaters holding hands and spinning, then they both rotate about a common point that lies halfway between the two of them. But now let's replace one of the adults with a child. The adult has a much greater mass than the child so when they spin around the child travels almost in an orbit about the adult, covering much more distance than the adult. But the adult is not stationary; seen from above the adult and the child will both still be seen to spin about a common point between them, albeit closer to the adult – this is the barycentre that was mentioned earlier. The adult will be wobbling around this common point due to the mass of the child. This second scenario illustrates how the Earth–Moon system works, where the child is the less massive Moon and the adult is the more massive Earth.

Now, to explain the second tide I am going the make things really weird. Let's say that my adult skater is wearing a rather strange costume, a doughnut-shaped rubber ring full of water. Can you picture this? I know it sounds odd. As the child and my strangely dressed adult spin around each other, the water in the rubber doughnut starts to bulge outwards, away from the child. This happens due to a rotational force known as the centrifugal force: when an object is moving in a curved path, this is an apparent force that pulls the rotating object away from the centre of rotation (in this case, the barycentre). In the instance of our skaters, that is in a direction away from the child, and in our Earth–Moon system in a direction away from the Moon.

So, on one hand, we have the gravitational force of the Moon pulling water towards it, but on the other hand, we have the centrifugal force from the rotation of the Earth pulling water away from the spin axis, which happens to be away from the Moon. So for our tidal system, we have two bulges of water and hence two high tides. And at 90 degrees to this, we get two low tides.

NOT TO SCALE

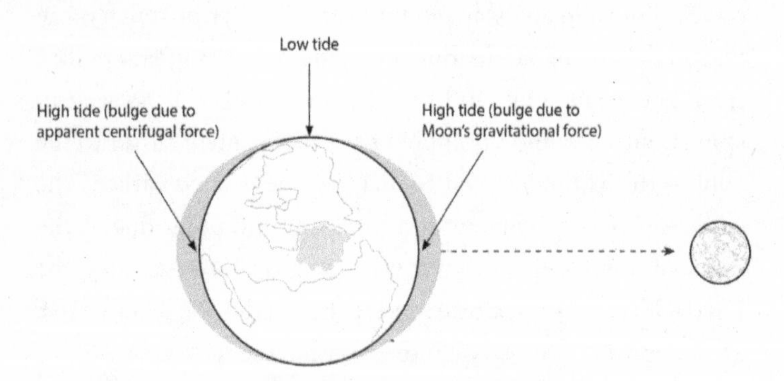

Moon's gravitational force and apparent centrifugal force causing two high tides.

OK, things got a bit strange there, but stay with me because things are going to get stranger.

I mentioned earlier that it is not just the Moon that causes the tides, the Sun does too. The Sun is far more massive than the Moon (nearly 30 million times more massive), but it sits a lot further away from the Earth (nearly 400 times further away). You might think that with this great distance the Sun's influence on the Earth's tides would be diminished, but this is not the case. In fact, the gravitational pull the Sun exerts on the Earth is about 180 times stronger than that of the Moon on the Earth (the Sun really is that massive). However, the Sun's effect on the tides is less strong – indeed, it's only about 46 per cent of that of the Moon. I know, even as I write it, it doesn't seem right, so to understand why the Moon dominates the tide we need to look a bit closer.

As I mentioned earlier, the fact is that it is not just the size of the force that matters but the difference experienced from one side of the Earth to the other: the differential force.

In the example of my ice skaters, there was a nice, simple system, but the fact is that the gravitational pull on the Earth by the Moon is not even. The closer you are to the Moon, the greater the pull, so one side of the Earth experiences a greater gravitational effect than the other (see the section on the inverse square law below, pages 116–17).

With the Sun it is just the same: the side closest to the Sun experiences a greater force. But with the Sun the difference experienced between one side and the other is much less, only a 0.02 per cent difference, whereas with the Moon the difference is 7 per cent; hence the Moon's effect on the tide is about 46 per cent stronger than the Sun's.

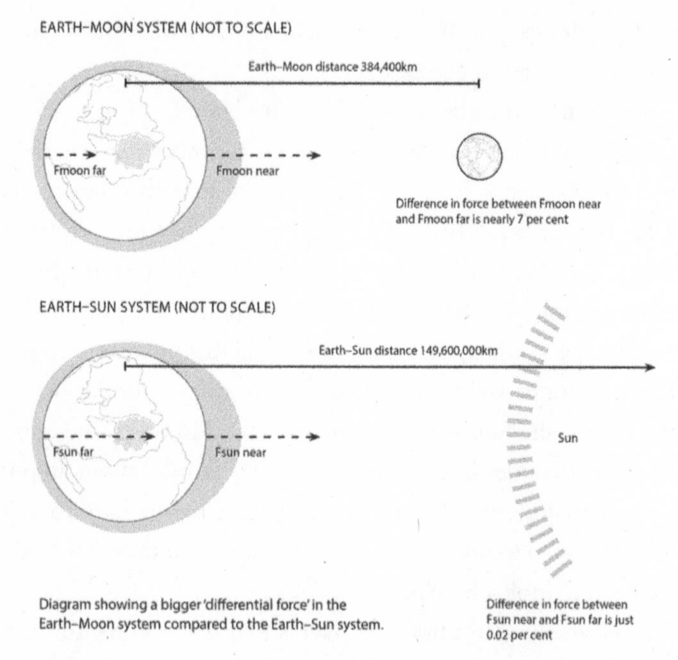

Diagram showing greater 'differential force' in the Earth–Moon system compared to the Earth–Sun system measured from the centre of the Earth to the centre of the Moon and Sun.

Things get interesting when the Sun and Moon are pulling in the same direction to cause the tides, enhancing the pull. This occurs twice every month when there is a full moon or a new moon. When this happens, you get a super high tide because the overall gravitational force of both the Sun and Moon pulling in one direction is greater than usual. It also occurs when the Sun and Moon are opposite each other in the case of a full moon, but in this case they are working together on the two opposite bulges. These large tides are called spring tides.

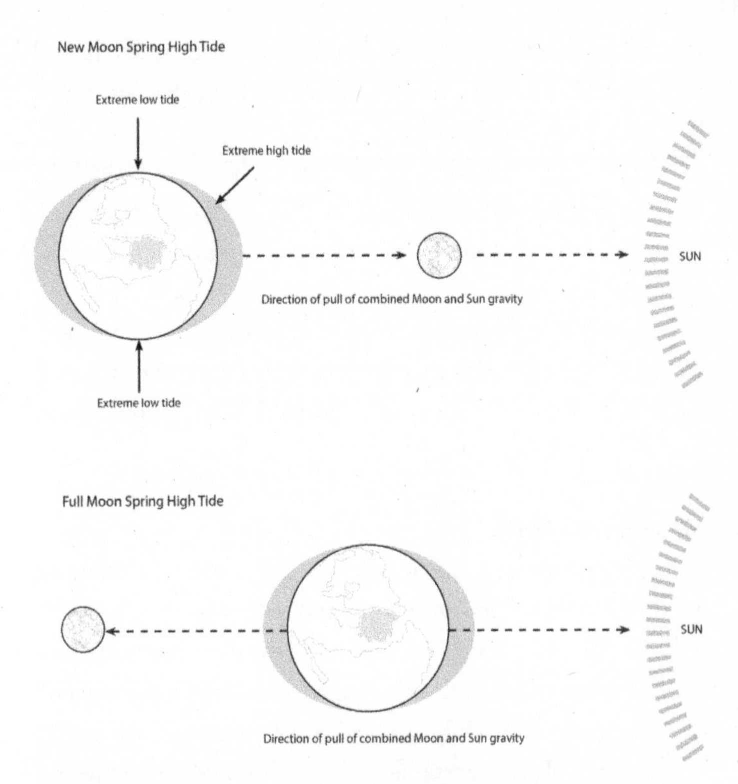

Extreme high tides occur when the overall gravitational forces of the Sun and the Moon are pulling in the same or opposite directions.

The opposite of this effect occurs when the lunar and solar forces are acting at right angles to each other. In this case we see shallower tides called neap tides.

There are other factors that can affect the tides too, such as the lay of the land in terms of topography. Wind and weather patterns also can affect water level: strong offshore winds can move water away from coastlines, exaggerating low tides, whereas onshore winds can push water onto the shore, making low tides much less noticeable. In addition, high-pressure and low-pressure weather systems can cause tides that are much higher or lower than predicted.

On average, the tidal bulge is about 1m high, but at the Falls of Lora in the west of Scotland, because of the geography of the coastline, it is four times higher, causing our spectacular phenomenon.

So that's tides explained in a rather large nutshell, but now we have the basics let's play around with the system. What if the Earth–Moon distance was different? What if the Moon was closer or farther away?

Moon on the Move

Rather than torturing our ice skaters any more, let's think up another scenario. Both magnetism and gravity work in a similar way: the closer two bodies are, the greater the force between them. So let's replace our skaters with a bag of iron filings and a magnet, or even better a ferrofluid (a liquid that becomes magnetised in the presence of a magnetic field). The ferrofluid represents the water molecules in the world's oceans and the magnet represents the Moon. When I move the position of the magnet nearer to the ferrofluid, the force is greater and the liquid bulges toward the magnet. But as I move the magnet away, the bulge gets smaller.

This is a basic law of physics, which explains the behaviour of both gravity and magnetism. It takes the form of the 'inverse square law', and essentially it means that the closer any two bodies are, the stronger the magnetic attraction – or, for our Earth–Moon system, the gravitational attraction.

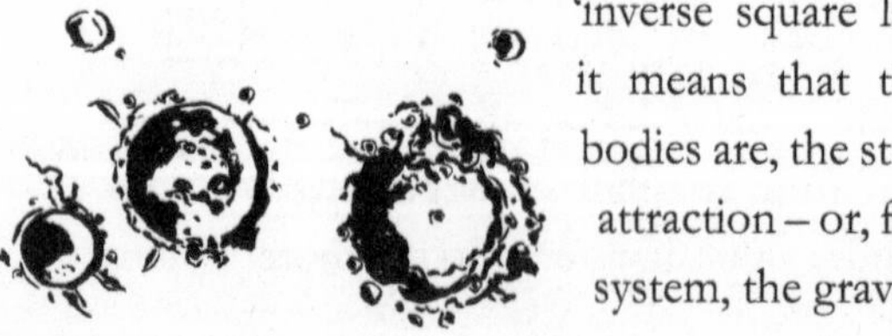

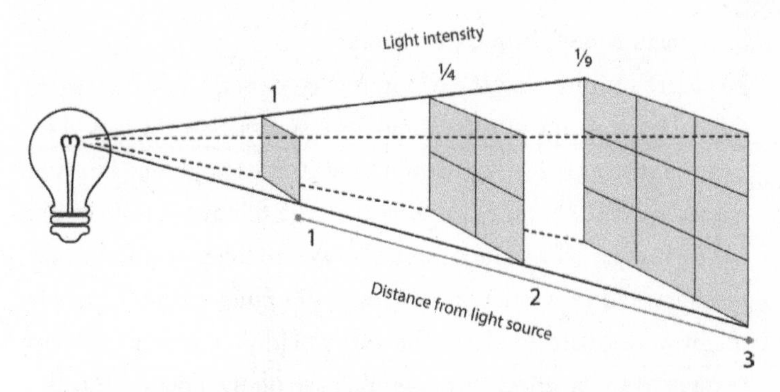

The intensity of light is inversely proportional to the square of the distance from the light source, so it is much weaker further away from the source.

But what if we moved the Moon 10 times closer, or even 20 times closer – what sort of tides would that produce?

To understand the law better, let's consider a point of light shining in a dark room. The intensity of light radiating from the source is inversely proportional to the *square* of the distance from the source. That basically just means that light becomes very much weaker further away from the source of the light. If you compare the intensity at one metre and two metres, you get a quarter of the light intensity at two metres than you do at one metre. And, at three times the distance, the light per unit area is down to one-ninth.

So what would this mean in reality if the Moon were twice as close? Well, due to the inverse square law, the gravitational force would be four times as strong and, around the world, tides would be higher and waves would be bigger. On the west coast of Scotland, the Falls of Lora would be monstrous, and low-lying coastlines would be flooded. And that is if the Moon were just two times closer!

Supersize Moon, Supersize Tides

Now let's imagine that the Moon is even closer, so that it was 20 times bigger in the sky. It would look glorious. Night would be closer to day as the Moon would be 400 times brighter. (The Sun is actually 400,000 times brighter than the full moon so it would not be like daylight.) The contrast between a full moon and a new moon would be vast. With a Moon 400 times brighter, people might find it hard to sleep. But this would be a minor problem compared to the effect that it would have on the tides of Earth.

At low tide, the waters would pull back and reveal great swathes of seabed, exposing all sorts of creatures. At high tide, a super tidal wave would rush in, engulfing all it encountered. As it moves over field and dale, through towns and cities, low-lying areas of land would be submerged. Holland would be covered in water, and the islands of Japan and New Zealand would have only their highest regions exposed. In the UK, the West Midlands would be the only part of the British Isles above the water line at high tide. Cities like London, Tokyo and New York would be wholly submerged, with just a few of the tallest buildings sticking out a little through the water. A huge, beautiful Moon carries a real sense of menace.

This crazy scenario seems like a far-fetched Hollywood disaster movie, but once upon a time, when the Moon was first formed, it really was this close to Earth and unleashed this sort of tidal power.

The Stuff of Life

The motion of the tides is the most familiar phenomenon we can see that has been caused by the Moon's relationship with Earth. But there is another, even more fundamental way in which the Moon has affected our planet. Recent research seems to indicate that it was the formation of the Moon that enabled the prime conditions for life to begin on Earth. But how did it begin?

Since we have been able to consider the question, humans have wondered about the origin of life here on Earth. What kick-started evolution into action? Why and how did the first single-cell organisms evolve?

To answer these questions we need to travel, at least in our minds, to a terrain that is similar to that of the early Earth. A volcanic landscape with smoking fumaroles and hissing mud pots, and lava seeping from cracks in the crust. Not a great place to live! But we need to picture a landscape like this to envisage the turbulence on Earth that occurred after the collision with Theia or another body that created the Moon, assuming that theory is correct (see page 49). Supersize tides followed the impact and the huge volcanic eruptions reset the atmospheric composition of the Earth in what turned out to be a very useful way.

New gases reached the surface in the form of free hydrogen,* hydrogen sulphide and methane. Today, we think of these gases as unpleasant and toxic, but for the early Earth, they were the stuff of life itself.

In the 1950s, the famous Miller–Urey experiment took place. It was named after the two chemists, Stanley Miller and Harold Urey, who were the first to conduct it. They tried to recreate conditions found on the early Earth by mixing together the three aforementioned gases and then heating them up to the high temperatures that would have existed at that time. They also added sparks (electrical discharge) to simulate the effects of early lightning.

What emerged from their experiment was a film of brown slime. It wasn't too exciting to look at, but what they had created was no less than one of the building blocks of life. Using this set-up they had created amino acids, the basis of proteins.

This was an amazing breakthrough discovery: it proved that it was possible to create the chemicals of life using nothing more than the elements that were present on the early Earth. But, on its own, this was not enough to explain the existence of life on our planet. A vital component was missing. How was information passed down from one generation to the next?

For any life form to evolve, it needs a way of passing on genetic information. Each protein needs to be able to make an

* Hydrogen on Earth is usually found in a molecular state, bonded to another hydrogen atom and forming a hydrogen molecule. With free hydrogen, the atoms remain unbonded, which means they are very reactive.

accurate copy of itself to pass on to the new generations. This is where a little-known chemical called RNA comes in.

RNA stands for ribonucleic acid and is the less familiar cousin of DNA. But just because it is less well known does not mean that it is not as important. RNA is the messenger in cells that allows DNA to communicate with the body's protein factories. And, crucially, it is self-catalysing – that is, it contains its own genetic information, so it can reproduce itself spontaneously.

As a result, many biochemists believe that it is the precursor to DNA – in other words, it is the most basic raw material of all life. But how did it suddenly emerge on planet Earth?

Emerging from the Tides

Over the years, scientists have tried to work out how to produce RNA using nothing but the elements and conditions that existed on the early Earth. In recent years, the scientists think that they have cracked it and now it seems that it's the Moon that plays a critical role in the process.

If we return back to the primordial Earth, some 500 million years after the formation of the Moon, things are starting to look more like the Earth we know today. The planet's surface has cooled down enough for there to be oceans of sulphurous water, separated by rocky outcrops of land. And as the Moon passes overhead, we see the coming and going of great tides. Don't forget that at this stage the Moon is sitting a lot closer to the Earth so the tides are huge.

Intertidal zones emerged: vast swathes of land that were exposed at low tide and then covered up at high tide. Go to a rocky beach or bay – the ones with little rock pools, full of a variety of creatures, dotted about the place – and you can visit these intertidal zones that were so important for the story of our evolution.

At every low tide, as the water receded, organic compounds got deposited in pools. As the day went on, the Sun rose and evaporated the water in these pools, so that the remaining chemicals became more and more concentrated, only to be churned up again with the next incoming tide. This ebb and flow, caused by the Moon's gravitational pull, was creating a rich chemical soup from which RNA could form.

But this is not just an abstract theory. This process can be recreated in the lab with some wet chemistry – the name for the type of chemical experiment you can do with liquid samples. Here is the recipe. Place your primordial chemicals in a conical flask. Heat them up and remove the evaporating water. This concentrates the chemicals. Rehydrate your chemical soup, repeating the process again. Keep up this cycle but also apply some ultraviolet light to simulate the intense UV radiation coming from the early Sun. As the process is repeated, a white liquid starts to appear. Analysis of this liquid will reveal that it contains strands of self-replicating RNA. We have created an essential building block for life.

When Charles Darwin wrote about the origins of life, he speculated that life had begun in a 'warm little pond'. Experiments such as these indicate that Darwin was more or less right. These intertidal zones were Darwin's warm ponds – primordial chemistry labs where the first life could evolve, all orchestrated rather beautifully by the passage of the Moon.

The Wandering Moon

So far the Moon has given us tides, the gases needed for life and the mechanism for the chemicals of life to reproduce themselves – quite a lot, by anyone's standards. But the Moon has done so much more than this.

Ever since life emerged 3.8 billion years ago, the Moon has continued to exert the most profound influence on evolution. How does it achieve this? It is because the Moon controls the speed at which the Earth spins.

I was given an insight into how this works when I was invited to visit one of America's largest telescopes, situated in the desert in New Mexico. It is at the Apache Point Observatory, the home of the Lunar Laser Ranging Project.

This project is one of the last vestiges of the Apollo programme. As mentioned, on some of the last human visits to the Moon, astronauts left behind retroreflectors, which are suitcase-sized reflector arrays positioned on the surface (see page 51). They reflect back beams of light that are fired from the Earth, so astronomers can measure the distance between the Earth and the Moon, with millimetre accuracy.

The actual process of taking the measurement from Earth is not easy. First, just as with most astronomy, you need a clear night. After conducting some visual checks with the telescope and sighting it on one of the retroreflectors (Apollo 15 when I was there), the bombardment towards the arrays begins. The Apache Point astronomers fire very intense pulses of laser light called photons up to the retroreflector. The return signal is minuscule. Of the 100 quadrillion (that is 100 million billion i.e. 10^{17}) photons per pulse that go up, only a handful return. It's a bit of a hit-and-miss process, once described as like trying to hit a moving target in a shooting gallery that is 3km away. But even with the return of just a few photons per pulse a distance can be calculated.

On the night of my visit we were lucky and a few thousand photons were pinged back to the telescope. By measuring the round-trip time, down to a trillionth of a second, it was possible to get a very precise measurement. The Moon's distance from us that night was 393,499km 257m and 798mm.

Now, my visit to Apache Point was quite brief but they have been taking similar measurements over the last 40 years and during this time they have acquired thousands of results. From these, they have been able to work out that the Moon is drifting away from us very slowly. Every year on average it retreats by 3.78cm. As I mentioned earlier, that is about the same rate as our fingernails grow.

So the Moon used to be closer but is now slowly drifting away from us. This long-term drift is happening due to the Moon speeding up (see box opposite), and as it speeds up it moves further away. But the problem is, as the Moon accelerates away from us, the Earth loses momentum and slows down. In other words, the spin, and hence the length of our day, is governed by the position of the Moon. The closer it is, the shorter the day, and as it drifts away from us, the longer our days get.

WHY IS THE MOON SPEEDING UP?

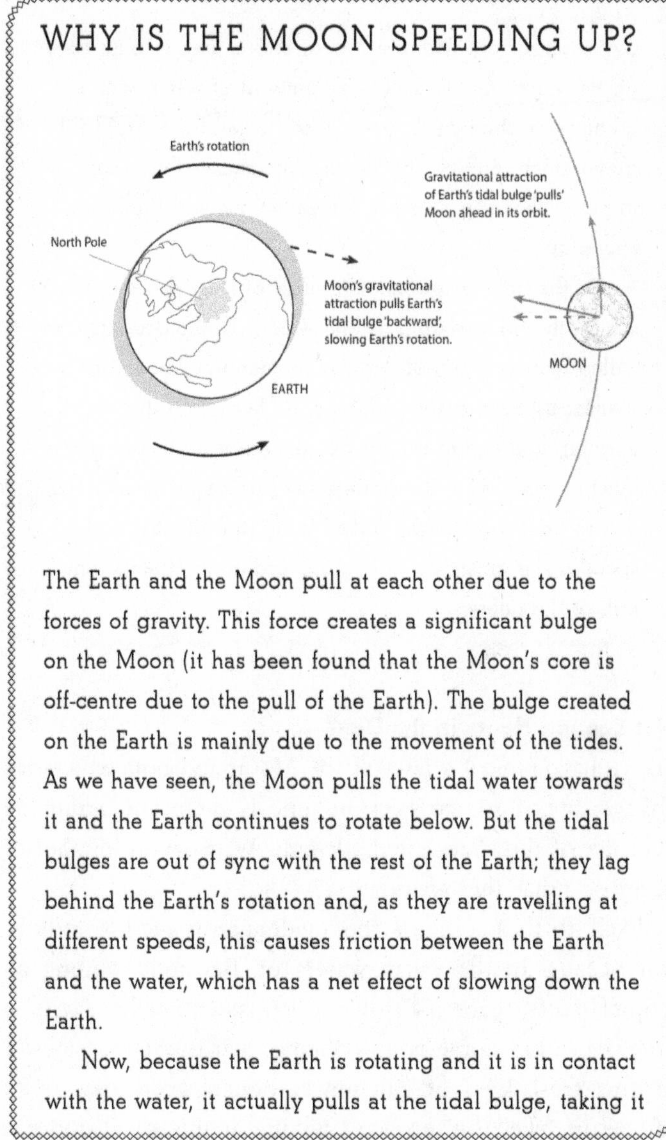

The Earth and the Moon pull at each other due to the forces of gravity. This force creates a significant bulge on the Moon (it has been found that the Moon's core is off-centre due to the pull of the Earth). The bulge created on the Earth is mainly due to the movement of the tides. As we have seen, the Moon pulls the tidal water towards it and the Earth continues to rotate below. But the tidal bulges are out of sync with the rest of the Earth; they lag behind the Earth's rotation and, as they are travelling at different speeds, this causes friction between the Earth and the water, which has a net effect of slowing down the Earth.

Now, because the Earth is rotating and it is in contact with the water, it actually pulls at the tidal bulge, taking it out of alignment with the Moon, and causes it to sit ahead of the Moon. As this is a large amount of water, with a great mass, the tidal bulge exerts a small but significant gravitational pull on the Moon. This causes the Moon to move ahead in its orbit, effectively making the Moon speed up.

So the tidal bulge has the effect of both slowing down the Earth and speeding up the Moon. This process is also called 'conservation of angular momentum', meaning that for a linked system such as the Earth and the Moon, angular momentum cannot be created or destroyed but must be preserved. So if one body (the Earth) slows down due to tidal friction, the linked body (the Moon) must speed up, in this case due to the additional gravitational pull of the bulge.

Not Enough Hours in the Day?

So we have pinged a laser at the Moon and noticed a small change, but if we go back in time is there any additional evidence of this change in day length? Were days really shorter in the past than they are now?

Well, to find some of this evidence we need to look in our oceans. In the warm waters of Bermuda, among its tropical reefs, there is a flower coral species called *Eusmilia fastigiata*. This coral is interesting because it secretes a distinct fresh layer of calcium carbonate every day of its life, as its 'skeleton' grows. Looking at this coral under a digital microscope and counting the layers, it is possible to work out how many days it has been alive. Because winter growth is slower than summer growth, it's also possible to discern the passing of individual years. It's just like counting the rings of a tree. So we can see how many days' growth happens within a year.

The same process can be conducted for a fossilised coral that lived 400 million years ago, during the Devonian period. By counting the fossil layers of day growth that occurred each year, it soon becomes clear that the days and years don't add up. Simply put, there are too many days in each year. Instead of 365, there are around 400 of them.

The absolute length of the year is dictated by our orbit of the Sun, which has been stable over billions of years. So, for there to be 400 days and nights in a year, there's only one explanation: the Earth must have been spinning faster back then and each day must have been shorter. Doing a bit of arithmetic, you can tot up the number of hours in a year (8,760) and divide it by the number of daily skeletal layers on the fossilised coral (400). It is possible to work out that in the Devonian period each day must have been just 21 hours 54 minutes long.

This seems to me to be an amazing thought. I have always assumed that days on Earth have always been 24 hours long. It feels like an astronomical standard, but the truth is that it is just the case right now, at this moment in the Moon's long drift away from the Earth.

So what happens if we go back even further? Doing some more sums, it's possible to work out that 4.5 billion years ago, when the Moon started its long journey, each day must have been much, much shorter: only 4.7 hours long. Instead of

365 days in a year, there were 1,827. Yes, the Sun would have risen and set 1,827 times each year and there would have been a sunrise and a sunset around every five hours.

To get a feel for such short days, there is only one place to go: the International Space Station (ISS) that has been orbiting the Earth since 1998 and playing host to humans from 17 different nations. To stay in orbit and not fall under the gravitational pull of the Earth, the ISS has to travel very quickly around the Earth at a speed of 27,000km/h. So each orbit of the Earth takes only 90 minutes. That means they get to see a fresh sunrise and sunset every hour and a half! That is a very short day. But what effect does this different day length have on the astronauts? Can their bodies adjust to 16 days and nights every 24 hours?

It's not an easy challenge. Nearly all animals, including humans, have a biological clock. We are programmed to eat, sleep, metabolise and regenerate cells according to a 'circadian rhythm', governed by the rotation of the planet on the current 24-hour cycle. And this process depends on visual cues, most importantly the rising and setting of the Sun. Our circadian-clock mechanism is embedded deep in the hypothalamus, a region of the brain, which controls the release of hormones. The mechanism has evolved over many millions of years to receive visual information through the eyes and secrete the feel-good hormone, melatonin. If the clock is disturbed, the hypothalamus produces the stress hormone, cortisol. Blood

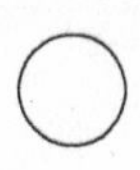

pressure rises and heart rates soar. So if these processes are so firmly engrained, how can astronauts on the ISS possibly cope with their day/night rhythm being so crazy?

One way to cope would be to avoid the problem altogether. Rather than the astronauts trying to make their bodies adjust to the intense ISS daily cycle, it might be easier to trick them into thinking they are still on a 24-hour cycle by using blackout screens and artificial lights. But, with a dawn and dusk occurring every 90 minutes, this would be a tall order, and virtually impossible to carry out effectively. Another problem would be that one of the reported joys of being up there in the ISS – and this is something that I would like to check out for myself, if anyone out there needs an astronaut – is the amazing view of Earth. Astronauts love to look back home; as Tim Peake put it, 'We always talk about seeing the view of planet Earth and how beautiful it is.' But this amazing view of home, of course, exposes them to the frequent sunrises and sunsets.

Rather than stopping them from looking out, the astronauts' vital signs are continuously and closely monitored, checking for any signs of stress. From the data tracked so far, it seems that for the most part the astronauts are able to cope with this bizarre reality. And it is precisely because it is so bizarre. Ninety-minute days are so very different and out of sync with our normal rhythms that the brain simply gives up trying to make sense of them, and stays more or less tied to a 24-hour cycle.

Ironically, it seems that it would be much harder to adjust if the day were only two or three hours different. For instance, if it were 21 hours long, the brain would try to adapt, but it wouldn't be able to cope. Many experts think we are unable to deal with anything more than a one-hour discrepancy either side of 24 hours.

Small Moon, Long Nights

This difficulty in adapting our circadian rhythm poses an interesting question. If we are so finely tuned to a 24-hour cycle that only a few hours' difference throws us, what will happen in the future as the Moon moves away from us and the days get longer?

Rather than waiting, let's imagine the world as we know it but with a more distant Moon. Let's take the night sky in a city like New York; suddenly the full moon changes and becomes smaller, the sky becomes darker. In places all over the world, people look up to see our same familiar full moon but noticeably smaller than normal. Even a small shift of just 10 per cent farther away – 38,000km – makes a dramatic difference (a shift of this size would still take around 1 billion years to happen).

The tides across the world are now weaker, and currents in the major rivers of the world are smaller. In the Thames, the Nile and the Hudson River, the variation of the tides is barely noticeable.

And the spin of the Earth has slowed to a crawl. Days and nights are now 12 times longer. The average day–night cycle is now 288 hours. That means that night time can last 144 hours.

If we were suddenly thrust into a change like this, how would different life forms adapt to cope with these super-long days and super-long nights? What would happen to our circadian rhythm? Would evolution take us on a different path?

Resetting the Clock

Again, this scenario sounds artificial but there are people living with this sort of challenging environment today. To meet them, we need to travel above the Arctic Circle. Here in northern Norway, in the Land of the Midnight Sun, people and animals really do have to cope with super-long days and nights. For three months, the sun barely rises. So, to start with, how do the animals in these areas cope?

At these high latitudes, any successful species has to adapt to the long days and nights. Some, such as bears and squirrels, resort to hibernation; they're able to drop their body temperature, reset their biological clock and conserve energy by sleeping for weeks at a time. This may be a useful strategy in our alternate world.

But reindeer seem to be utterly unfazed by the long days and nights of the Arctic. This is because they have evolved a way of decoupling their clock genes and bypassing their circadian rhythm. Instead, they catch sleep in short bursts when they feel like it – not when light levels dictate. It is called 'polyphasic' sleep.

Experiments have been conducted with bull reindeer fitted with an activity monitor. This enables scientists to watch the males' activities and get a minute-by-minute account of their sleep–feed balance. When analysed, it was found that reindeer have a three to four hour feeding cycle: they feed, ruminate (digest) and then feed again. And while they ruminate they sleep. Or at least they display all the hallmarks of slow-wave sleep, but they remain conscious and aware of their visual surroundings, including any potential predators.

Could other animals develop the same knack of decoupling from the day cycle? Cattle, sheep and goats, for instance, are all ruminants and they have a similar biology to reindeer. It would not be beyond the scope of imagination for them to head down the same evolutionary path.

But what about us humans? As primates, could we possibly evolve similar sleep patterns to the reindeer, if faced with super-long days and nights? To answer this question, it would appear that the perfect research subjects are located with the reindeer. They are the local Sami reindeer herders. For thousands of years, they have lived at high latitudes with the reindeer and would surely be ideal candidates to be tested to see if they had mastered the trick of polyphasic sleep. But unfortunately, it is 50 years too late!

The arrival of artificial lighting has changed the game. The Sami people are now able to bypass external cues and live more or less on a normal 24-hour cycle, no matter what external factors are in play. As research subjects, they would be tainted.

Thankfully there is another – rather small – group of people we can study for evidence of successful polyphasic sleep in humans: round-the-world solo sailors. These sailors have to be as alert as possible for 24 hours at a time, so they sleep during the day and night in one to two hour snatches, independent of visual cues, just like reindeer. This tentatively suggests that, if necessary, we humans do have the potential to reset our biological clocks.

But if the idea of living with a small, distant Moon and significantly longer days does not seem too scary, then I need to tell you about another rather more dramatic impact that a distant Moon would cause.

A World at Right Angles

So, here we are, with the Moon 10 per cent further away from us. The tides are weaker, days and nights are significantly longer ... but why should that turn a city like Las Vegas, usually a hot, sunny location, into a dark, arctic desert? The temperature in the city is now -20°C and the buildings are covered in snow and ice. What is going on?

To answer this question all we need is a ball and a steady hand. You may have seen the phenomenon on the internet of individuals known as ball wizards showing off their tricks. They can take a ball, set it spinning and then – seeming to defy gravity – they can make the ball balance on a fingertip.

This is possible because the spinning ball is more stable than one that is not spinning. Try the same trick with a stationary ball and it is game over, the ball just wobbles and falls off.

Our Earth, under the gravitational influence of the Moon, is remarkably stable – spinning on an axis that is not quite upright, but angled at a constant 23 degrees, within a degree or two. If there were no Moon – or even if the Moon were just 10 per cent further away – this stabilising effect of the fast spin would disappear. There would be nothing to stop our Earth from tipping over. The angle of the tip could fluctuate anywhere between 0 and 90 degrees. A tip of 90 degrees would mean that the poles would end up sitting where the equator used to be.

It sounds mad but we believe that this is one of the things that happened to Mars many millions of years ago. The result of that catastrophe meant that a planet that once had liquid water flowing over its surface became the dry, arid place we see today, with very little sign of liquid water present.

If the Earth did tip over by 90 degrees, the world's climate would go haywire. River systems would dry up; there would be torrential rain where today there is a desert. Pack ice would be common in the tropics.

Places above the tropics like Vegas would be dark and freezing all winter. They would then experience not one solstice (when the Sun is overhead) but two, in spring and autumn, while the summer would be relatively cool with the Sun lower in the sky. And in November, three months of darkness would return.

It would be a truly odd world, but the biggest impact would be at the poles. With three months of constant intense sunshine each year, the ice caps would melt. This would force a huge amount of fresh water into the world's oceans. Global sea levels would rise by 60m.

From space, we would see a very different picture of the world. The shape of the continents would have to be redrawn, huge areas of land would be submerged and every coastal city in the world would be gone!

This could easily be our destiny if – or rather when – the Moon's gravitational pull becomes too weak to keep us upright and stable. That's the bad news. The good news is that it will take another billion years before we reach this tipping point so we do have some time to prepare.

Why tell such a frightening tale of a future so far way? Well, the aim of thought experiments such as this is not to scare us but to make us appreciate how lucky we are that the Moon is exactly where it is right now. Small shifts in its position would have a real impact on life here, and our planet would potentially be a very different place. Things really are so fragile and so finely balanced.

So what has the Moon ever done for us? Well, apart from tides, setting up the environment to make life possible, allowing the creation of the building blocks of life, giving us our current 24 hours in a day and of course keeping the Earth's rotation stable, it also gives us some amazing sights.

I have already confessed to being a self-certified lunatic and I think that this might be true for you too. Just one look at the Moon grounds me and gives me peace, no matter what phase it is in. Whether it appears as an everyday crescent or in one of its more unusual and breathtaking forms, such as a full eclipse, it is full of inspirational beauty. And we can truly say it is the mother of us all.

OBSERVING THE MOON

Speaking of the Moon as beautiful and inspirational, this seems like a good time to discuss how best to observe it. No other astronomical body dominates our night skies the way the Moon does. And at certain times it is visible during the day too.

Most other astronomical bodies need some form of augmentation to appreciate them. The Sun is too bright to be viewed directly unfiltered and is therefore impossible to observe unaided. The stars can be appreciated as constellations but remain pinpoints of distant lights. The Moon, however, is the ideal subject for naked-eye observations.

Although the Moon has the benefit that it can be viewed in far from optimal conditions due to its intrinsic brightness, as with all astronomical observations there are a few simple steps that can make the time you spend outside observing it a lot more profitable and comfortable. Growing up in London, I cut my metaphorical astronomical teeth watching the Moon and not always in the best of conditions, and it is a pleasure that I still enjoy today. Here are a few tips to optimise what you can see without having to eat a couple of pounds of carrots every day (which apparently wouldn't help your eyesight anyway).

Ideal Viewing Conditions

The Moon is bright relative to other objects visible in the night sky, but there are things you can do to improve your ability to see the detail on the surface. The main factor that will affect what can be seen is light pollution. If you live in or near a city or large town, positioning yourself away from street lights is essential. This is often harder than it sounds, but if you can find a clearing such as a playing field or park that is less well lit, this is often a good start. Also, many big cities give off a glow of light, so even if you do not live within a city the light pollution from a neighbouring conurbation can still affect your viewing. If possible, observe away from the city glow, as it can overwhelm the light of the objects under observation.

When I was at university in central London, I would often go to the outskirts of Hyde Park to admire the firmament. However, do be aware of your personal safety when going into less well-lit areas and err on the side of caution; ideally, go stargazing with a friend or two, and always let someone know where you're going if it's slightly off the beaten track.

If an open area is not accessible, minimising the number of street lights visible in your line of sight will definitely help. At my home in Surrey, my back garden is relatively free of human-made light pollution, which is mainly emitted from the street lights situated at the front of the house. In parts of Surrey, we also have the benefit that between 12am and 5am the street lights are turned off as a cost-saving measure. This gives us much less light pollution and makes finding a good location significantly easier. As a child in London, I would often go to a small area between the council flats where we lived, where the street lamps were fewer.

Dark Adaptation

Before you start to observe, you need your eyes to be in a state of 'dark adaptation'. In bright light, the pupils of our eyes contract to let less light in, stopping us from being dazzled. If we then move to a dark area our pupils dilate, but this reaction takes some time. So to make the most of a night's viewing, we need to allow our eyes to adjust to the dark by letting our pupils fully dilate. As well as the pupils' response, much of our dark adaptation is dependent on the refresh rate of the chemicals in the light receptors located in the retina.*

The retina contains two types of light-sensitive cells called rods and cones. Rods work in low light levels and in black and white, whereas cones require much more light to function and give us colour vision. Both rods and cones contain chemicals that break down or bleach when exposed to light. Our sensitivity to light is dependent on the amount of the unbleached chemical that is present.

---

* This is the area at the back of the eyeball where the light-sensitive cells sit. It connects to the brain via the optic nerve.

Our eyes have two main modes of operation. In bright light, the cone receptors are mainly operational while, conversely, the light-sensitive chemical rhodopsin in our rods is bleached. When we encounter low light levels, the rhodopsin starts to regenerate but this can take some time. During this regeneration period, our eyes are not as sensitive to the darker conditions as they could be, and, as a result, we can see less detail. The adjustment to low light levels can take around 30 minutes, and during this time it is essential not to be exposed to bright light or the process of dark adaptation will be lost. While a torch is a useful tool for observing expeditions, care should be taken not to shine it in your own eyes or in anyone else's direction. Eyes that are dark adjusted can respond quite painfully to the full glare of a torch, and even if the flash is short it does take some time for the eyes to recover and see dimmer details again. Some torches are available with a red filter or, even better, a red LED. These are useful because they are less dazzling but can still provide some help in illuminating one's way.

Comfortable Viewing

People often seem to think that observing the Moon should be done standing up; this certainly can be the case, but a comfortable chair can be useful too. I find a deckchair ideal for when I am doing naked-eye observations. I set it up at a reclined angle so that I can sit in it with my head supported but looking up in the desired direction. As the Moon moves across the sky, the chair is light and relatively easy to reposition. I have a friend who uses his children's trampoline for observations. If the sides are not too high, you can get a good view of the night sky while being gently supported by the springs. I have been told that this is comfortable enough to lie on for hours.

THE MOON WITH THE NAKED EYE

Having found your optimum conditions for observing the Moon, what can we see on its surface with just the naked eye? Well, what is visible on the lunar surface will be very much dependent on the phase of the Moon – so let's start by looking at what that means.

Phases of the Moon

The Moon, when visible, is usually the brightest thing in the night sky. Nonetheless, it is important to note the vital role that the Sun plays in our observations of the Moon and of the solar system as a whole. The Sun is effectively the source of light that allows us to observe other objects around us. The Moon and the planets do not generate any light of their own, so to see them we need to detect the sunlight reflected off their surfaces.

Just as with the Earth, half of the Moon's surface (the day side) is always illuminated by the Sun. However, as the Moon orbits the Earth, the sunlit portion that can be observed varies through the course of the month as the Moon moves around us.

The various views or phases of the Moon have been divided into eight handy descriptions that give us an understanding of how much of the daylight side of the Moon is visible from our viewpoint on Earth, as the month progresses.

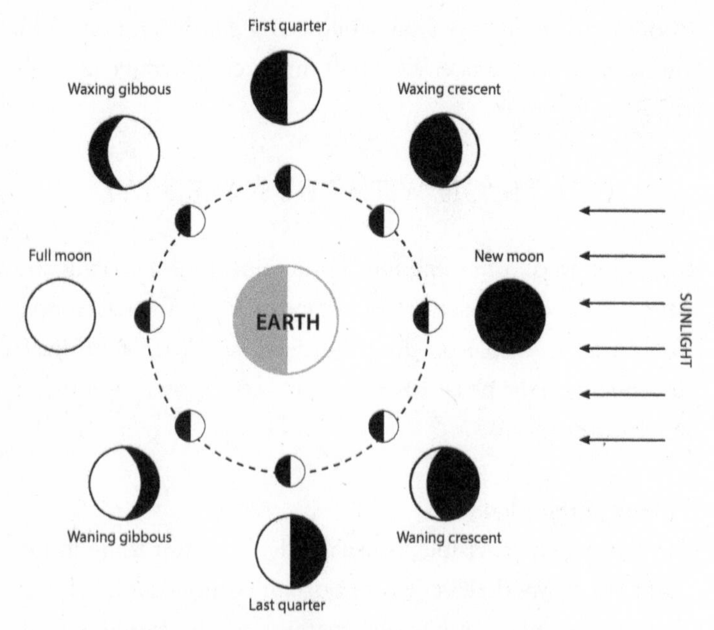

The 8 Phases of the Moon.

The first phase for us to consider is the 'new moon'. Strangely enough, this is when the Moon's illuminated side is not visible from Earth at all, because the Moon is sitting between the Earth and the Sun. So a new moon is actually invisible to us. Occasionally, if the alignment of the Sun, Moon and Earth is just right, then a total eclipse of the Sun by the Moon is possible (see pages 156–7 for why don't we have eclipses every month). A total solar eclipse can only occur during the new-moon phase of the lunar cycle.

The second phase of the Moon is called the 'waxing crescent'. Waxing in this instance means gradually increasing in magnitude or extent, which sums up nicely what the Moon is doing. During this period the Moon is getting brighter and also more of it is luminated and visible, going from what we call in our house the 'thin Cheshire Cat' smile from *Alice's Adventures in Wonderland* to the next phase, the first quarter.

The 'first quarter' initially seems like a misnomer as half of the Moon's disc is visible at this point. But if we take into account the Moon's true spherical nature, then indeed a quarter of the Moon is visible from Earth due to light from the Sun. This occurs approximately one week after the new moon.

The next phase is called the 'waxing gibbous'. The term gibbous comes from the Latin word *gibbus*, meaning humped or hunched, and went on to mean something rounded or protuberant. This aptly describes this phase of the Moon, where more than half of its sunlit side can be seen.

With the Moon waxing with time we eventually get to the 'full moon' approximately two weeks after the new moon. This is the Moon at its most striking, with its daylight side fully visible from Earth. The Moon now sits on the opposite side of the Earth from the Sun in a position called 'opposition', meaning it is on the other side of the sky. A total eclipse of the Moon can only occur in this phase.

Having reached maximum brightness, the only way from here is a diminishing moon, which is termed 'waning'. This comes from the old English word *wonian*, which means to lessen or diminish. This phase is called the 'waning gibbous'. It looks similar to the waxing gibbous with more than half of the daylight side of the Moon visible, but the luminated visible area of the Moon is diminishing as the cycle goes on.

The 'last quarter' is the next phase of the cycle when, as with the first quarter, half of the Moon's lit disc is visible from Earth.

The final phase is called the 'waning crescent' and it occurs when a crescent Moon is visible, but it is waning and growing smaller as the cycle proceeds.

Sometimes people also like to talk of the 'old moon', a very thin crescent that occurs just before the Moon is invisible again at the start of the cycle, the new moon.

WHY DO WE SEE THE MOON DURING DAYLIGHT?

The phase of the Moon that I most like to see is the waxing crescent, but with the new moon being invisible and the UK skies often being cloudy, it frequently feels as if I have not got my Moon fix for a number of days. Then at around sunset I see the waxing crescent in the darkening skies and I feel much better. For people with keen eyesight, it is possible to see the Moon in this phase during the day too.

The table below gives an indication of where the Moon is in the sky relative to the Sun in the northern hemisphere. So, for instance, my favourite, the waxing crescent, can be seen in the dark sky for three hours after sunset, at which point it sets itself. As it is trailing behind the Sun it is possible to see it in the daytime sky too. Three hours after sunrise this phase will rise above the horizon but may be hard to spot due to the sunlight. At the other end of the scale, the waning crescent leads the Sun by three hours so you may be able to spot it most easily early in the lightening skies of dawn. Once the Sun has risen it will still be possible to see it, but, again, it's much harder to spot.

Moon phase	When visible relative to Sun
New Moon	0 hrs
Waxing Crescent	TRAILS Sun by 3 hrs
First Quarter	TRAILS Sun by 6 hrs
Waxing Gibbous	TRAILS Sun by 9 hrs
Full Moon	TRAILS or LEADS Sun by 12 hrs
Waning Gibbous	LEADS Sun by 9 hrs
Third Quarter	LEADS Sun by 6 hrs
Waning Crescent	LEADS Sun by 3 hrs
New Moon	0 hrs

The High Seas

When observing the Moon with the naked eye, the easiest things to spot are the lunar maria. These dark lunar 'seas' are quite visible and show up in good contrast to the lighter highland or terra areas. The lunar maria were mainly named in 1651 by the Italian priest/astronomer Giambattista Riccioli, whose lunar naming system has been taken as standard. Another astronomer, Johannes Hevelius, suggested a different naming system at the same time but Riccioli's system stuck, probably due to the evocative and romantic language he used to describe the seas and oceans.

Some of the maria that stick out from the crowd are:

Sea of Rains (Mare Imbrium): this is one of the larger maria but it is not quite an ocean. The sea itself has a diameter of 1,596km (991 miles). Current research suggests that it was made after a cataclysmic impact with a small protoplanet in the Moon's past.

Sea of Serenity (Mare Serenitatis): this mare has a diameter of 674km (419 miles) and is one of the locations of a mascon (see page 39). Both the Soviet Luna 21 and the US Apollo 17 landed in close proximity to this mare.

Sea of Crises (Mare Crisium): some 555km (345 miles) in diameter, the strangely named Sea of Crises is another location where a mascon has been found. The Soviet space probe Luna 15 crash-landed at this site in 1969 and the Luna 24 mission returned a sample of lunar regolith from this location in 1976.

Sea of Fertility (Mare Fecunditatis): this mare is 840km (522 miles) in diameter but so far has been found to be mascon-free. The Soviet Luna 16 returned the first lunar sample from here in 1976.

Sea of Nectar (Mare Nectaris): only 340km (211 miles) in diameter, the Sea of Nectar is one of the smaller maria but it is darker in colour, which makes it easier to spot with the naked eye.

Sea of Clouds (Mare Nubium): another evocatively named mare, this one is 715km (444 miles) in diameter. Scientists from Spain observed and recorded an impact in this crater in September 1993.

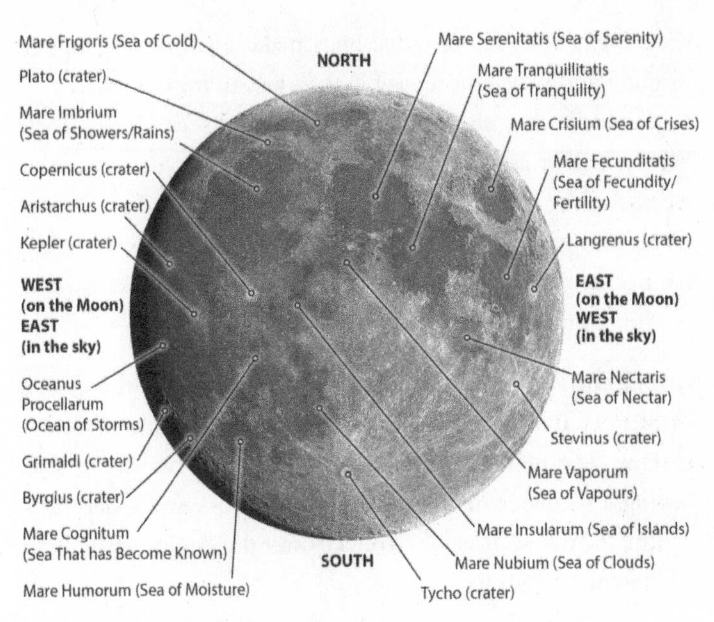

Map of the Moon's craters and maria.

Ocean of Storms (Oceanus Procellarum): bigger than the average lunar mare, this has a diameter of 2,592km (1,611 miles). The area is easily spotted with the naked eye because of its sheer size. The manned mission Apollo 12 landed here and a number of other unmanned Soviet and US missions.

Contemplating Craters

If you are lucky enough to have exceptional viewing conditions then some of the Moon's more prominent craters may be visible too. Most of the craters visible to the naked eye have two things in common. Firstly, they have bright rays radiating out from them. This is ejecta material that was thrown out during their formation. These radial lines make them appear bigger, brighter and hence more visible than just an impact crater.

Secondly, they sit in maria. Impact craters are usually fairly light in colour as they are relatively new and have not been as weathered by the solar wind as other areas. A bright impact crater sitting in a dark mare is more easily spotted with the naked eye because of the contrast between the colours of the two.

Some of the more visible craters are:

Aristarchus: this is a prominent crater that sits in the Ocean of Storms. It is one of the brightest formations on the Moon's surface. The crater is larger than the Grand Canyon in size, having a diameter of 40km (25 miles) across and a depth of 3.7km (2.2 miles). Riccioli named it after the Greek astronomer Aristarchus of Samos.

Copernicus: this crater is about 100km (62 miles) in diameter, and has an extensive system of rays. This crater looks at its best when viewed close to the terminator (the division between the illuminated and dark hemispheres of the Moon – see pages 162–3) when the prominent rim is lit up against the contrasting shadow of the crater floor.

Kepler: this bright crater, like its near neighbour Copernicus, has a system of rays. Unlike most features, which are best seen at or near the terminator, the rays are most visible when the Moon is full. Its diameter is approximately 32km (20 miles).

Tycho: this crater sits near the south pole region of the near side of the Moon. It is pleasingly circular in shape and is surrounded by bright ejecta radiating out from it. It has a fresh look about it and is thought to be relatively young as its rim and ejecta area seem not to have been pummelled by lunar impacts. It has a diameter of 85km (52 miles).

The trio of Aristarchus, Copernicus and Kepler can be seen well into the last quarter, but the best time to observe them is two to three days on either side of a full Moon.

Earthshine: Ghost Moon

As well as spotting features on the lunar surface, an interesting thing to look out for when observing the Moon with the naked eye is a phenomenon called earthshine. It appears as a pale glow that lights up the unlit portion of the Moon, like a ghostly version of a full moon.

Earthshine is most clearly visible a few days before and after a new moon when the crescent moon is close to the horizon at sunset or sunrise. In the past, people looked at this spectacle and wondered what was happening. They called the phenomena 'ashen glow', or said that 'the old moon is in the new moon's arms'.

This astronomical marvel was eventually explained by the polymath Leonardo da Vinci, who as well as all his other achievements had a keen interest in astronomy. He realised that as well as light leaving the Sun, hitting the Moon and being reflected back to the viewer on Earth, some of the sunlight could also hit Earth, be reflected up to the Moon (some of it hitting the night side of the Moon) and then get re-reflected back to an observer on Earth. The Earth light, hitting the night side of the Moon, is what causes the earthshine.

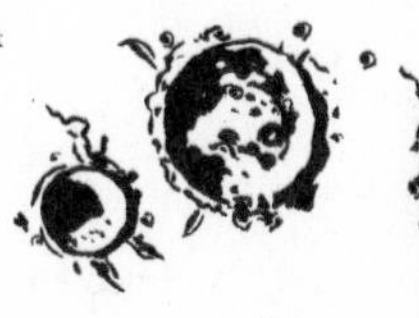

Total Solar Eclipses: A Cosmic Coincidence

A total eclipse of the Sun occurs simply when the Moon gets between the Earth and the Sun, and blocks the sunlight reaching a certain part of the Earth. It sounds quite mundane but it is a truly glorious spectacle.

The beauty of a total solar eclipse occurs due to a truly cosmic coincidence. It happens that, seen from Earth, the Sun is 400 times the diameter of the Moon, but the Moon is 400 times closer to Earth, so as viewed from our perspective the Sun and Moon look as if they are the same size in the sky. It means that the Moon can perfectly cover the Sun, giving us the majesty of a total eclipse.

The fully shaded area of the Earth or Moon experiencing a total eclipse is called the 'umbra', from the Latin for shade or shadow, and the area that is experiencing a partial eclipse is called the 'penumbra'. The 'antumbra' is the area where you can see an unbroken ring of light around an eclipsing planet or moon, known as an annular eclipse. All these types of solar eclipses are discussed here, and all of them are both rare and beautiful.

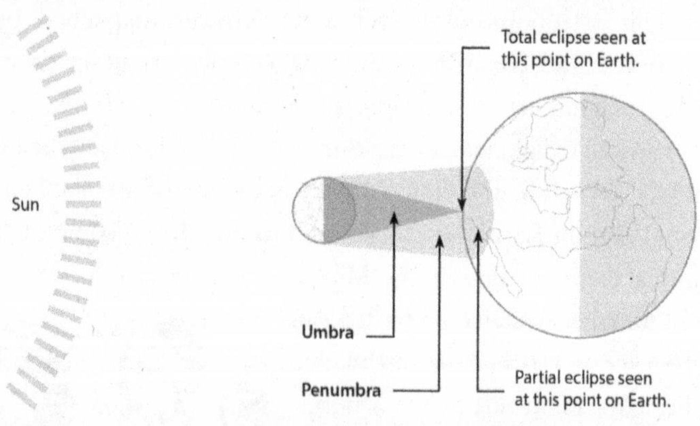

A total and partial solar eclipse.

I was able to see a total eclipse myself in 1999 when I travelled out to Le Havre in France, which was in the area where the total solar eclipse was viewable. It was one of the most memorable events of my life. I was very lucky as it was a cloudy day but the skies cleared about ten minutes before the eclipse so I could see it clearly with my solar-eclipse glasses. It began when the light started dimming to something like twilight; birds started to prepare to roost as the ambient light got dimmer and finally there was the moment of totality when the Moon passed precisely between the Earth and the Sun. At this moment you get a brilliant halo of sunshine visible around a dark Moon and the Sun's outer layer, called the corona, can clearly be seen. It is possible to see prominences (spikes of solar material) flowing off the Sun's surface. Totality lasted for just a few minutes and, as the Moon appeared to move on, I saw the brilliant diamond-ring effect when the first rays of sunlight shone past the Moon before the usual daytime brightness returned. It was truly awe-inspiring and I remember having to sit down for a few minutes afterwards to recover as my senses were so blown away by the proceedings.

Having enjoyed the first one so much I then travelled to the US to see the 'Great American Eclipse' of 2017 and, as it occurred in August during the school holidays, we went as a family. I can still remember the awe on my daughter's face, lit just by the Sun's corona. Truly magical.

On a more practical note, it is important to be aware that any observations of the Sun, aided or unaided, should always be performed with suitable eye protection. Directly looking at the Sun can seriously damage your eyes.

Do bear in mind that the Moon is slowly moving away from us, so the beauty that is a total solar eclipse will be lost to future generations. Instead, we will see annular eclipses – one of the four types of solar eclipse we can observe today.

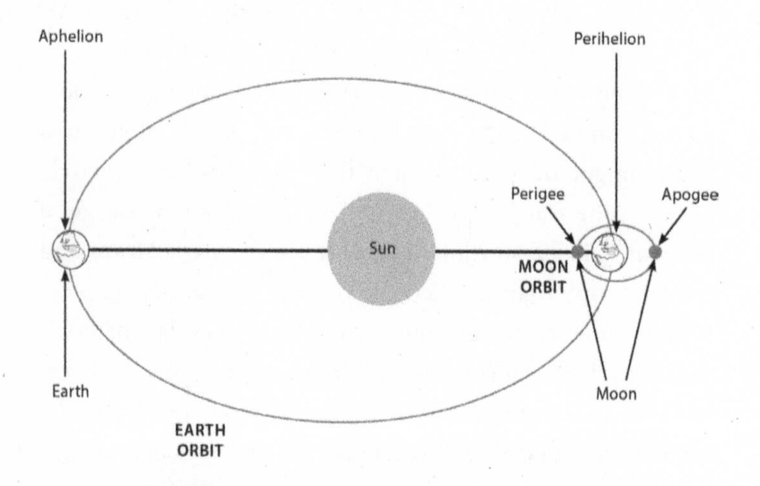

Diagram showing the elliptical orbit of the Earth, including the points at which it is furthest away and closest to the Sun (the aphelion and the perihelion). Similarly, the Moon's orbit around the Earth is elliptical. The point where it is furthest away from us is the apogee, and where it is nearest is the perigee.

In the diagram here we can see a very elliptical orbit. In contrast to the total solar eclipse when the Sun is totally obscured by the Moon, an annular eclipse occurs when the Moon is at apogee, i.e. in its micromoon state, further away from the Earth, meaning it looks smaller and does not fully cover the diameter of the Sun. The visible outer edges of the Sun form a 'ring of fire' or annulus around the dark Moon.

Partial solar eclipse

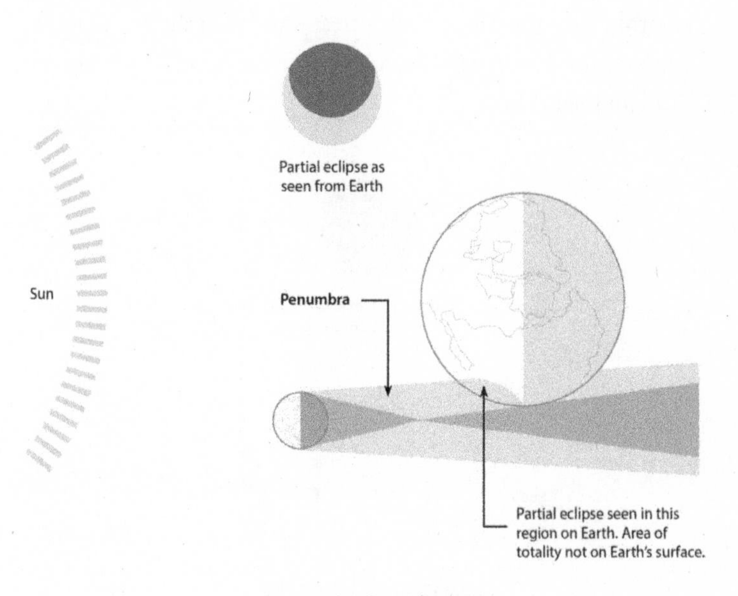

A partial solar eclipse.

A partial eclipse occurs when the Moon comes between the Earth and the Sun but is not aligned in a perfectly straight line. In this case, the Sun's disc is partially covered by the Moon but a significant proportion of the Sun may still be seen. In the diagram above, the region of totality, i.e. the area from where you would experience a total eclipse, is not on the Earth's surface.

Annular solar eclipse

Many of us have heard of a supermoon when the Moon appears larger in the sky; well, the opposite of a supermoon is a micromoon when it looks smaller in the sky. These variations in size are caused by the Moon's elliptical orbit. If a total solar eclipse occurs when the Moon is in a micromoon position in its orbit, then the Moon will appear smaller in the sky than the Sun and the Moon will not fully cover the Sun. This is known as an annular eclipse.

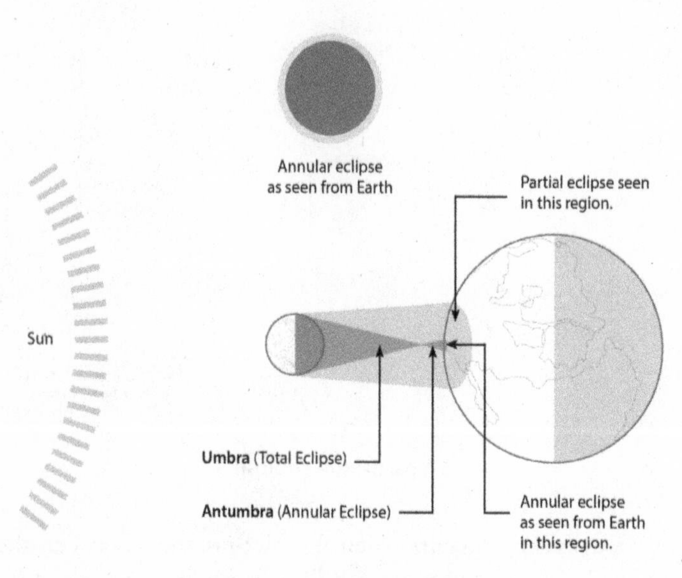

An annular solar eclipse.

Hybrid solar eclipse

This form of eclipse is very rare. It occurs when an eclipse changes from being an annular solar eclipse to a total solar eclipse as the Moon moves along its orbit.

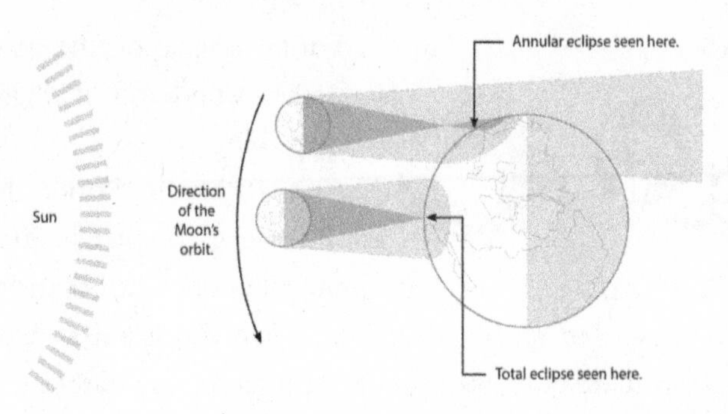

A hybrid solar eclipse.

Lunar Eclipses: Strange and Eerie

Just as there are solar eclipses, lunar eclipses can occur too. This happens when the Earth gets fully or partially between the Sun and the Moon, throwing the Moon into shadow. As the Earth is much larger than the Moon, a total lunar eclipse lasts much longer than the few minutes of a total solar eclipse, often lasting a few hours.

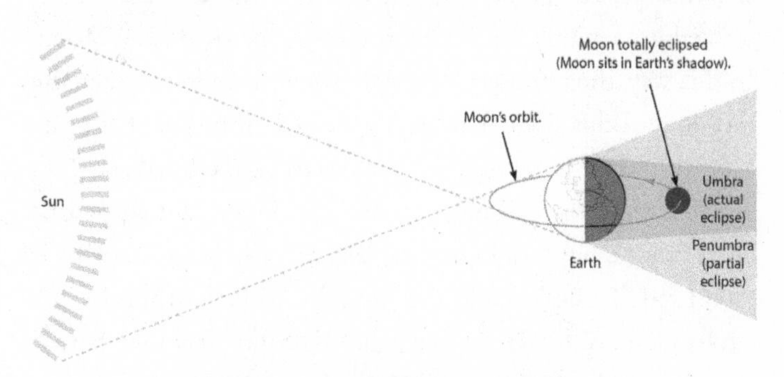

A total eclipse of the Moon, where the Moon sits in the shadow of the Earth.

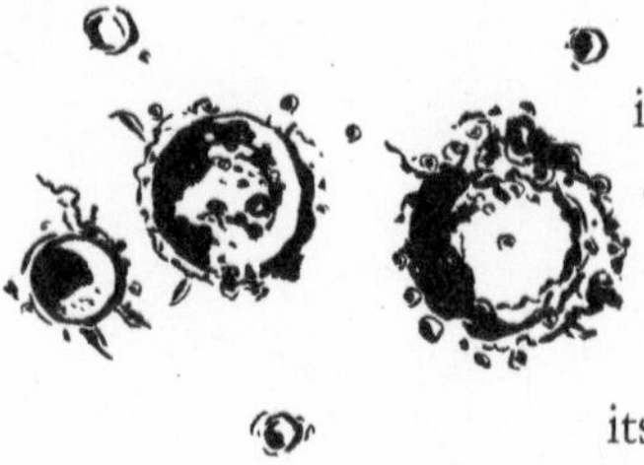

A total eclipse of the Moon is a rather wonderful, if slightly creepy thing to see. In the Earth's shadow, the Moon does not disappear completely due to its limited illumination, but instead it goes a deep red, often described as blood red. If you are not expecting it, it can come as a rather nasty surprise and I can imagine people of the past being what I can only describe as 'well freaked out' on seeing this. It gave me a nasty turn on my first viewing, filling me with a strange feeling of foreboding – although I knew in scientific terms what was happening.

It occurs due to an atmospheric phenomenon called scattering. In the shadow of the Earth, you would expect little sunlight to reach the Moon as the light is effectively blocked by the bulk of our planet. But as we know, the Earth has an atmosphere surrounding it, and this can act like a lens, refracting/bending the sunlight towards the shadowed Moon. In this way, the atmosphere allows some sunlight to reach the Moon, and this light then gets reflected from the shadowed Moon's surface to be beamed back to us on the Earth.

This explains why we can see the Moon during a total lunar eclipse, but it does not explain why the Moon's colour is blood red. To understand this we need to look at the Earth's atmosphere. Although we see light originating from the Sun as a warm orangey/yellow, this light is in actual fact made up of a range of different colours, called a spectrum.

It is thought that the Sun's spectrum was first seen and understood by Sir Isaac Newton in the seventeenth century, when he passed a beam of sunlight through a prism and discovered a rainbow of colours emerging from the other side.

Indeed, the rainbows we see projected across the sky are one of the ways we can see the different colour components of the Sun's spectrum. Newton's prism was a barrier that the light needed to pass through, and as different colours (or wavelengths) of light travel through glass at different speeds, the output of his prism was multicoloured. In the case of the rainbow, drops of rain act as the barriers rather than a prism, and as the sunlight passes through these drops, the different colours are again impeded to different extents and hence they emerge from the drops separated by colour.

But in a total lunar eclipse, the Moon goes red, not rainbow-coloured. This is caused by atmospheric particles. The light does not pass through these particles, but it is scattered off them instead. But in a similar way to the prism, different colours of light are scattered to different extents.

In our atmosphere, there are two types of this phenomenon, called Mie and Rayleigh scattering. The form of scattering that occurs is dependent on the size of the particles involved and the wavelength/colour of the light that is being scattered. The molecules that make up our atmosphere (mainly nitrogen and oxygen) scatter light via Mie scattering, but they scatter shorter wavelength blue light more than longer wavelength red light.

The long path that the sunlight takes from the Sun to the Moon via the Earth's atmosphere means that much of the blue wavelength light is scattered out. This means that the light reaching the Moon via this route is mainly red.

Hence the Moon looks deep red during a total lunar eclipse. These scattering phenomena also explain why sunsets look red and why the sky is blue.

But what about the 'blue Moon' – is it really ever blue? We hear about the blue Moon in songs, and rare events are described as happening 'once in a blue Moon', but in all my time Moon watching I have seen the Moon in many guises – blood red, yellow, orange – but never the colour blue.

Surprisingly, a blue Moon has nothing to do with a colour change or a scattering phenomenon, but relates to what seems to be an 'extra' moon in the calendar. In some years, there are 13 full moons rather than 12. It is thought that the word 'blue', in this instance, is a corruption of the old word 'belewe', which meant 'betray' and reflected the strangeness of having an extra moon. Since 1946 (when an article in *Sky & Telescope* magazine defined it), the term blue Moon has also been used when a single month has two full moons. This usually occurs in January when the full moon is very early in the month, so there is another one on or before 31 January. However, if January has two full moons, then it is followed by a February with no full moon. If this occurs, then March will have two full moons. So a blue Moon is rare, but not that rare. They are a bit like buses: you wait years for one, then two come along at once.

WHY AREN'T THERE MORE ECLIPSES?

While we're on the subject of eclipses, there is one question that arises quite often, which is: why don't we get a solar and lunar eclipse every month?

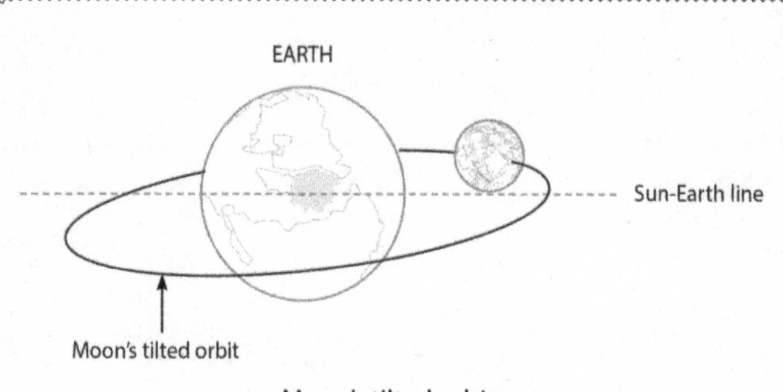

Moon's tilted orbit.

Well, if we draw a long line between the Sun and the Earth, then another one between the Earth and the Moon, we would see that the two lines would not be parallel. In fact, one would be tilted by about 5 degrees to the other.

This misalignment means that most of the time, the Moon's shadow or the Earth's shadow is too high or too low to be seen. But occasionally the alignment works and the shadow is projected at just the right angle to hit the other body – and that's when we get an eclipse.

Supermoons

Supermoons have featured a lot in the news of recent years. They can happen at the time of a full moon or a new moon and they make the Moon, if visible, look larger than average due to its closer proximity.

As we saw above, a supermoon occurs because the Moon's orbit about the Earth is not quite circular, as we usually assume, but an ellipse (more like a squashed circle).

Elliptical orbit of Moon around Earth.

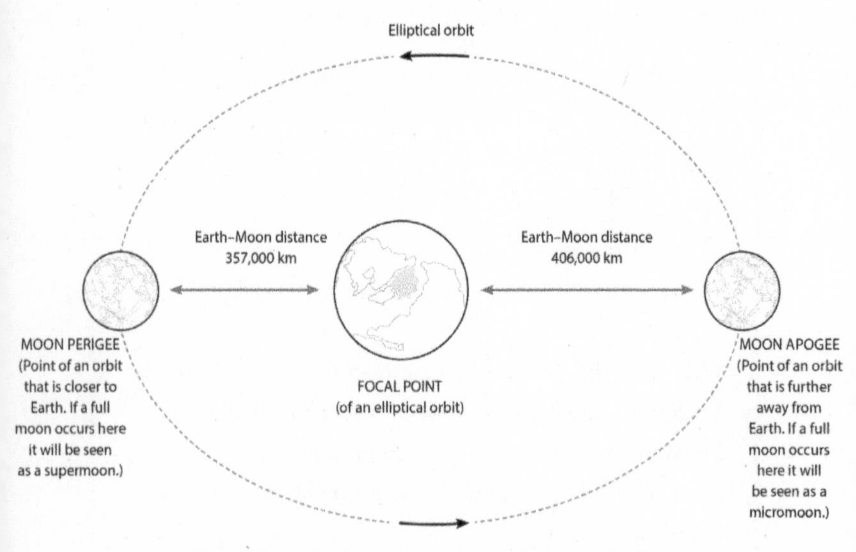

The elliptical orbit of the Moon around the Earth.

The perigee is the point in the Moon's orbit when it is closest to the Earth. When it is at this point the Moon is nearly 50,000km closer to the Earth than when it is at apogee. If a full moon occurs when the Moon is close to perigee it is termed a supermoon. The Moon, being closer to the Earth, will look bigger – up to 14 per cent bigger and 30 per cent brighter than a full moon occurring at apogee. A full moon occurring at apogee is called a micromoon, but we don't hear so much about them.

The image opposite shows the difference in size between a supermoon, an average Moon and a micromoon.

Interestingly, 'supermoon' is not actually a scientific term but was originally coined by an astrologer, Richard Nolle, in

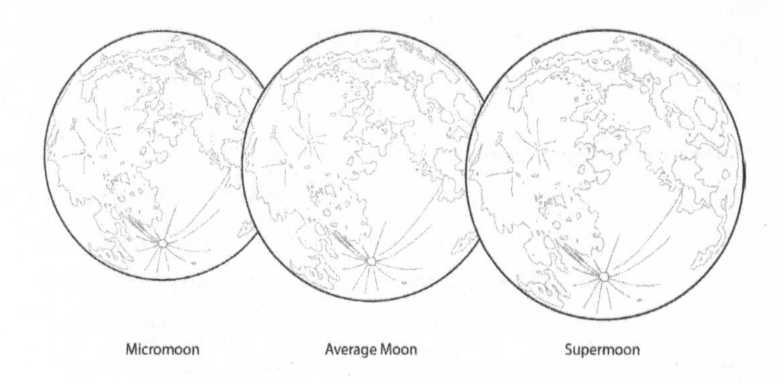

1979. He defined it as 'a *new* or a *full moon* that occurs when the *Moon* is at or near (within 90 per cent of) its closest approach to Earth in its orbit'.

The astrophysicist Neil deGrasse Tyson sent out the following tweet regarding the use of the term supermoon in 2014:

> @neiltyson: July's full moon is to August's 'Super Moon' what a 16.0 inch pizza is to a 16.1 inch pizza. I'm just saying.

Here he was comparing a supermoon with an average Moon. However, he did go on to say:

> @neiltyson Enthusiasm, interest, people going out and looking is good. I'm willing to pay this particular 'hype' cost. And it was pretty.

Whatever its origin, it looks like the term supermoon is with us to stay. And considering that the technical term for this phenomenon is 'perigee-syzygy',* I think that's the best choice, unless you happen to be playing Scrabble.

Low in Sky, Big Moon

One of my favourite times for observing the Moon, from a purely aesthetic point of view, is when it is nearly fully, waxing or waning and low on the horizon. Like this, it looks enormous, much larger than when viewed at a higher angle in the sky. If you have noticed this too, do not be alarmed. This is not like the supermoon; the Moon's size has not changed and has not moved by any significant amount. The effect is thought to be the result of an optical illusion.

Mario Ponzo in 1911 demonstrated that the mind judges the size of an object based on its background. To see it in its most simple form, draw two slanting but near-vertical lines on a piece of paper, like railway tracks heading up the page. Between these lines draw another two short horizontal lines of the same length as each other, one above the other. This is probably best done with a ruler to minimise cheating. The upper line generally looks longer than the lower line, even though we know they're the same length. It seems that a similar effect happens when the Moon is observed low in the sky. We compare it with objects that we are familiar with – trees, buildings, mountains and the like – and in this context the Moon looks large. However, when it has travelled higher in the sky there are fewer points of reference so its size appears diminished. If you are finding this idea hard to swallow then try this simple experiment. Hold a coin up to a low-lying Moon and compare its size to the coin.

---

* 'Syzygy' is a term used to describe a conjunction or alignment of three bodies, so in the case of a full or new moon it is the Sun, Moon and Earth.

As you trace the trajectory of the Moon through the night sky, the relative size of the Moon to the coin will stay the same.

LUNAR HIGHLIGHTS WITH BINOCULARS

All of the above lunar spectacles can be seen with just the naked eye and each of them has its own special magic. But at some point you may want to observe the Moon in more detail. Often when people get the astronomy bug and want to take that next step into augmented observations, the temptation is to go out and buy some great kit – which usually takes the form of a telescope. However, for Moon observations, and while you are still finding your astronomical feet, a good pair of binoculars can be a wonderful starting point.

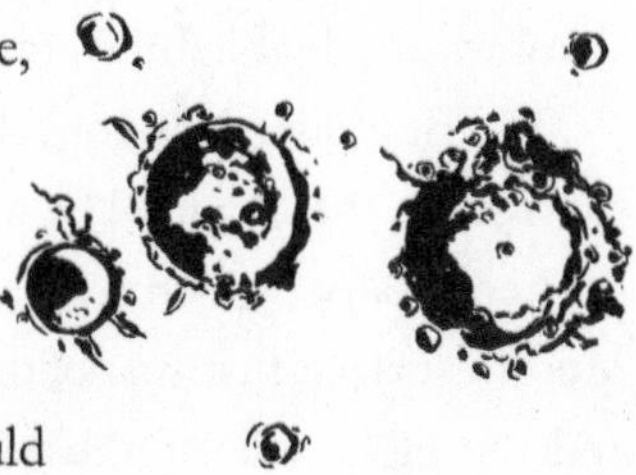

The main advantage is that they are very easy to use. Viewing the Moon through a telescope can take some practice: it tracks across the sky faster than other further-away objects, so unless the scope has a good tracking system you will be limited to the lower end of the magnification scale. Anything else will need a constant readjustment of the telescope as the detail you have focused on will drift out of view. Binoculars also have the added benefit of being very quick to set up. If the conditions are changing it makes sense to have a quick look with a set of binoculars rather than taking time to set up a telescope, only to find the conditions have changed for the worse. A good set of astronomical binoculars is also usually much cheaper than a telescope. I would recommend them to anyone taking their next steps after naked-eye observing.

For stargazing, astronomical binoculars are needed as they have a much larger aperture (light-gathering opening) than standard garden or bird-watching binoculars. However, because the Moon is the brightest thing in the night sky, if you look at it with some standard binoculars they will still give you a boost in observation compared with the naked eye.

So you have a clear night, you have your binoculars ... what should you look at first?

Well, just as with naked-eye moongazing, the phase of the Moon plays an important role in what can be observed. You might think that the full moon with its maximum illumination would be the best time to observe it; after all, half the Moon's surface is lit up for viewing. But as counterintuitive as it sounds, this is not the case. While it is less critical for naked-eye observations, viewing the Moon with binoculars or a telescope is generally best done when the Moon is not full.

During a full moon, the Sun's rays are hitting the lunar surface head on (at a perpendicular angle). When viewed under these conditions much of the glorious lunar detail is lost in the harsh glare of this light. During other phases, when we are observing light hitting the Moon at a glancing angle, more detail is visible as the lunar features are thrown into relief with shadows and highlighted areas.

The first quarter is considered by many to be the best time to observe the Moon. The area to focus on is a region called the 'terminator', which is the line of shadow that runs along the Moon's surface that marks the boundary between the daylight and the night side of the Moon. Along this line, the Moon's

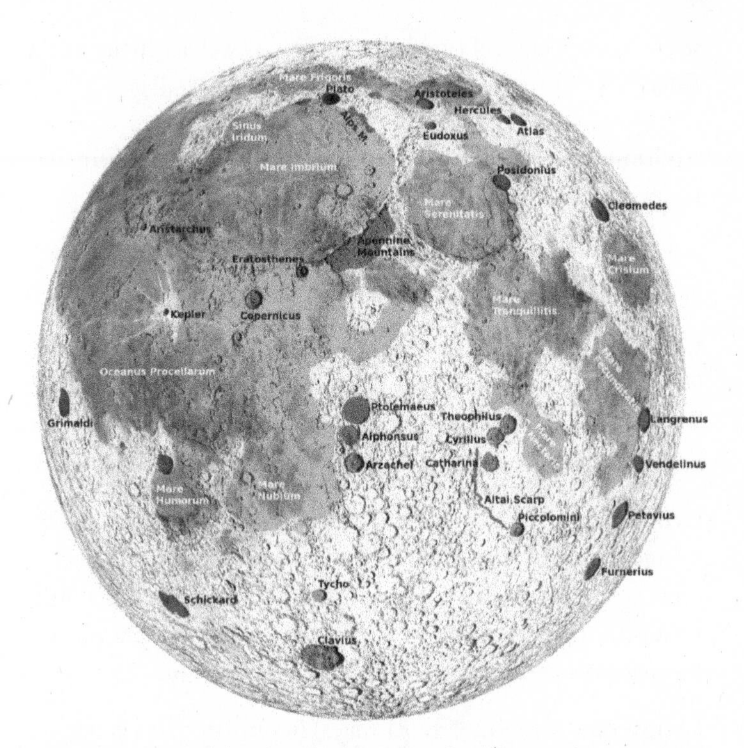

This is the level of detail on the Moon's surface that can be seen on a fine night with a good set of binoculars.

features are thrown into relief with the greatest contrast between the lit and the shadowed areas. The above sketch shows the level of detail on the Moon's surface that should be visible to an observer with a set of astronomical binoculars on a fine night.

Additional Sights to See with Binoculars

In the section previous (see pages 143–7), I included a list of features on the Moon that can be observed with just the naked eye. Here I would like to add a few additional features that a set of binoculars will enable us to view, as well as giving us an enhanced view of the features visible to the naked eye.

Archimedes: this crater is about 82km (51 miles) in diameter. It sits on the eastern edge of the Sea of Showers.

Aristoteles: this is another impact crater that is 87km (54 miles) in diameter. It sits in the northern area of the Sea of Cold.

Clavius: at 225km (140 miles) in diameter, this is one of the largest and oldest craters on the Moon, thought to be around 4 billion years old. It can be located using the distinctive Tycho crater mentioned earlier, as Clavius sits directly south of this.

Grimaldi: on the extreme western edge of the Moon, this is a large crater, 174km (108 miles) in diameter, which makes a great contrast with the exceptionally bright Aristarchus a little further north. Grimaldi is one of the darkest craters on the Moon, and very easy to spot when the Moon is full.

Langrenus: some 130km (81 miles) in diameter, this is one of the most prominent craters to become visible on the waxing crescent moon. It lies on the eastern shore of Mare Fecunditatis.

Plato: at about 100km (62 miles) in diameter, this is one of the more easily distinguishable craters because of its extreme northerly location and its distinctive dark hue.

LUNAR HIGHLIGHTS WITH AN AMATEUR TELESCOPE

When observing the Moon, a good map is an essential navigational tool to give you an understanding of what you're seeing. The sketch on page 165 gives an indication of the level of detail that can be observed using a good amateur telescope.

This level of detail can be observed with a telescope.

Magnification

With the use of a telescope you suddenly have access to a much wider range of magnification than you can get with binoculars. Choose your magnification depending on what detail you want to see.

If you go for X50, for example, this will give you a view of the whole Moon, which may enable you to get your bearings. A magnification of X150, by contrast, will enable you to identify many of the Moon's features. Just as we can star-hop from constellation to constellation, crater-hopping along the terminator can be good fun too. It's amazing to see the variety in size and shape of the different craters, and along the terminator they can be seen at their best.

The only time to avoid the higher magnifications is when the Moon is low on the horizon. When it is in this position you are observing the Moon through a thicker layer of atmosphere compared to when observing upwards (at zenith). The increased atmospheric depth means more turbulence and this, coupled with higher magnification, will result in an image that will be dancing in your viewfinder.

Photography

One of the pitfalls of most astrophotography is the lack of light available to capture. This is rarely the case for the Moon, which means photography is a lot easier.

Complex systems to mount a camera on to your telescope are available but for some simple snaps a smartphone can replace your eye at the eyepiece – some quite decent images of the lunar surface can be taken using this simple method.

Mountain Ranges

So far, we have just looked at the maria and craters that can be viewed on the Moon. With a telescope, you can pick out other details – for example, a number of mountain ranges can be seen on the lunar surface. Here is a list of some of the best ones to observe with a small telescope:

Taurus Mountains (Montes Taurus): far over to the east is the landing place of the last of the manned lunar explorers, Apollo 17.

Jura Mountains (Montes Jura): all around the edge of the 'Bay of Rainbows' crater is a mountain range generated by the same impact. These are the Jura Mountains, and this ring of mountains on the Moon is one of the most visually attractive.

Apennine Mountains (Montes Apenninus): named after the Italian mountains, they are one of the largest and most prominent mountain chains on the lunar surface. They are around 600km (370 miles) across and nearly 5km (3.1 miles) high. It is thought that they were created when the Sea of Rains (Mare Imbrium) was formed some 3.9 billion years ago.

Two of the most impressive mountain peaks on the lunar surface sit on the floor of the Mare Imbrium. They are called **Mons Piton**, which rises to a height of 2,250m (7,380 feet) and **Mons Pico**, which reaches a height of 2,400m (7,870 feet).

Another mountain range in this region is the **Caucasus Mountains** (Montes Caucasus). It is virtually a continuation of the Apennines to the north-east.

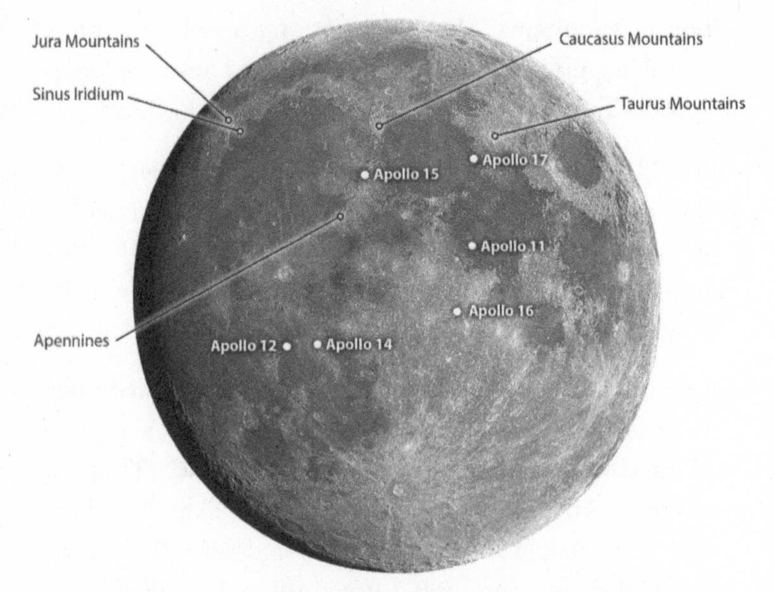

Map showing Apollo landing sites (by mission number) and mountains on the Moon.

The Apollo Landing Sites

Another highlight afforded by a telescope is looking at the Apollo landing sites on the Moon. Although no detail can be discerned unless you are using a top-end space telescope, it is possible to locate lunar features near to these landing sites:

- **Apollo 11**: landed in the Sea of Tranquillity (Mare Tranquillitatis), 20 July 1969.
- **Apollo 12**: landed in the Ocean of Storms (Oceanus Procellarum), 19 November 1969.
- **Apollo 14**: landed in the Fra Mauro Highlands, 5 February 1971.

 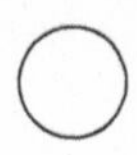

- **Apollo 15**: landed in the Sea of Rains (Mare Imbrium), 30 July 1971.
- **Apollo 16**: landed in the Descartes Highlands, 21 April 1972.
- **Apollo 17**: landed in the Taurus Mountains, 11 December 1972.

Looking at these landing sites leads us rather nicely on to our next section. So far we have covered our ability to study the Moon using the naked eye, binoculars and, finally, telescopes. But what was the next step for mankind in terms of getting closer to the Moon? Well, as we know, it was actually more of a giant leap …

REACHING FOR THE MOON: THE MOON LANDINGS AND THE SPACE RACE

In the 'Moon Past' chapter we saw how observations of the Moon from here on Earth enabled us to gain an understanding of that heavenly body. But the twentieth century enabled us to take the next steps in understanding the Moon by actually, physically going there.

The culmination of all this activity happened on 20 July 1969, an epic moment in history. The world stopped still to watch the first people leave the gravitational confines of Earth and land on the celestial body that we had observed from afar for so long. Images of this amazing giant leap for mankind were beamed across the globe.

Today, approximately 50 years on from that world-changing moment, it is interesting to analyse what led up to that point in time, and what we have learned and achieved since in terms of our relationship with our nearest neighbour and cosmic companion, the Moon. How did we get there and what impact did that moment have on us all?

To understand the environment of the Moon landings we need to understand the time in which they occurred and the events that led up to them.

The space era was born out of a very dark past. War is a great stimulus for the development of technology. The First World War led to the advanced development of the flamethrower, chemical weapons and the machine gun. The Second World War, with the ongoing advances in technology, led to the development of the long-range ballistic missile under the direction of a young graduate student called Wernher von Braun. Von Braun caught the astronomy bug as a young child after his mother gave him a telescope as a confirmation present. He was inspired by the land-speed records achieved by rocket-powered cars and there are stories that, as a child, he tried to recreate these feats by attaching fireworks to his go-cart. The results were not quite as desired and von Braun ended up detained by the police until his father came to collect him. Yet from this early experimentation, von Braun went on to get a PhD in aerospace engineering from the University of Berlin. The public title of his thesis was 'About Combustion Tests', but the real title which was kept classified by the

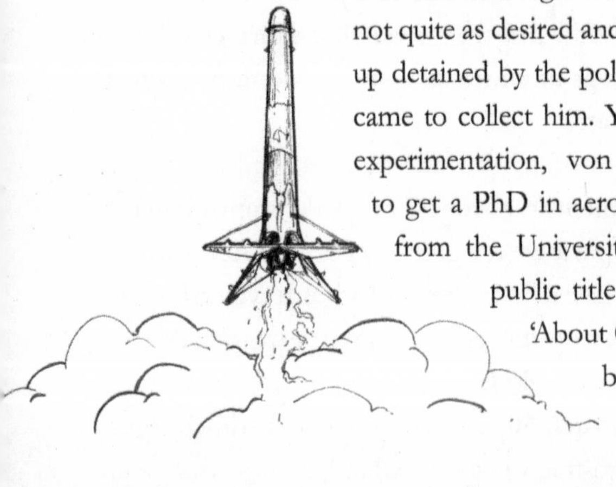

German army was: 'Construction, Theoretical, and Experimental Solution to the Problem of the Liquid Propellant Rocket'.

Indeed, von Braun and his team, using work pioneered by the American Robert H. Goddard, were making liquid-propelled rockets. By 1934, his group had made two liquid-fuelled rockets which had reached altitudes of 2km. These were the predecessors of the German army's death machines, named 'vengeance weapons', designed to attack Allied cities as retaliation for the bombings of German cities. This type of rocket, later called the V-2, was the forerunner of the rockets that we use for space exploration today. V-2s had the ability to carry warheads over great distances, so that they could be launched in France or Germany and rain terror on the major cities of Europe. This was a much more efficient way of delivering destruction than aircraft carrying bombs. With rocket technology you could launch annihilation from the confines of your own country without engaging the enemy head on.

The V-2s were constructed by concentration-camp prisoners at the Mittelwerk site in central Germany. When the bombs first hit London, the British government tried to conceal the development of the new German threat by blaming the explosion on a gas leak. However, when Germany announced the deployment of the new weapon, Winston Churchill went on record and admitted that London had been attacked by rocket-propelled bombs for a number of weeks.

Having no real defence against this new form of weaponry, and in response to the immense destruction being caused by the V-2s, British intelligence arranged for misleading reports to be intercepted by the Germans, suggesting that the V-2s were overshooting their London target by up to 32km (20 miles). In a recalibration effort, the German army adjusted the trajectories of the rockets, causing them to fall short of London and hit the much less densely populated areas of Kent. British intelligence then sent out reports stating that the bombs were now hitting their marks with maximum devastation. From September 1944, over 3,000 V-2s were launched by the German army at targets across Europe.

Significantly, the V-2, as well as being the first long-range ballistic missile, was also the first object to reach outer space. In tests on 20 June 1944, a V-2 rocket was launched that reached a maximum altitude of 176km (109 miles). The interface between the top of the Earth's atmosphere and outer space is considered to be 100km (62 miles) above sea level. So this V-2 far exceeded the Earth's stratosphere.

The 100km limit is not a cut-off point between space and Earth as such; the Hungarian-American engineer and physicist Theodore von Kármán is thought to have been the first to work out that above this altitude the atmosphere is so thin that an aircraft would need to travel faster than the planet's orbital velocity to have enough lift to stay in the air. The imaginary boundary is named the Kármán line in his honour.

With rocket technology having made its first foray into space, countries across the world wanted access to this new form of weaponry. In 1945, the Second World War was ended by the deployment of US atomic bombs on the Japanese cities of Hiroshima and Nagasaki. The German military, in the final weeks of its decline, decided that rather than the enemy gaining the advantages of its research, the best protection would be to destroy all the details of the V-2 programme and all those who had worked on it.

Both the Russians and the Americans moved in quickly to salvage what they could of the project and its personnel, but the Americans got there first and succeeded in securing von Braun and his team and shipping them out to the US. The Second World War was over but the Space Race had just begun.

The year is now 1957, and the first artificial satellite is launched from Baikonur in the USSR. This feat was the culmination of the work of an individual called Sergei Korolev and his team of scientists and engineers. Korolev was little known to the world during his lifetime as the USSR kept his identity a closely guarded secret until his death in 1966.

As the USSR progressed to putting the first human into space, Yuri Gagarin, on 12 April 1961, the Soviets were increasing their early lead in the race. However, later that same year, the then American President John F. Kennedy made a promise to get the first people to the Moon's surface before the end of the decade.

At the time that Kennedy made this promise, the USA had only put one person into space and even then he had not gone into orbit but had just undertaken a sub-orbital flight* that had lasted a mere 15 minutes. To go from this short flight to putting people on the Moon required a huge leap in technology and a lot of money. In the end, the Americans achieved their goal in just eight years.

How did they do it? When Kennedy made his promise no one had any idea of how they were going to get people to the Moon. And to achieve their goal many things had to fall into place.

The first crucial factor, of course, was money. To develop the technology quickly, the USA had to spend up to 4 per cent of its federal budget on space technology during the Apollo years.

Popular support for the mission was also key – to spend the huge amounts of money necessary, the American people needed to be happy with the project going ahead. As the USSR seemed to be winning the Space Race, public opinion was in favour of the project. It was clear that the USSR had its eye on the Moon too with its Luna missions, and the politics of the time meant that the Communists were considered the new enemy. Winning this next stage of the Space Race felt like a matter of safety as well as national pride.

The final critical factor was, of course, the technological development itself. To go from a sub-orbital flight to a Moon landing, the Americans needed to make major developments in several areas. To get all the equipment into space, they developed the Saturn V rocket, the largest space rocket ever

---

*A sub-orbital flight occurs when the object launched makes a very simple trajectory into space but does not go into orbit about the body.

built. Landers had to be developed to get people to the Moon's surface safely, along with spacesuits that were strong enough to protect the astronauts from the Moon's hostile environment but also flexible enough for the person inside to move easily. Countless challenges had to be overcome.

The USA finally achieved their dream on 20 July 1969 when Neil Armstrong, Buzz Aldrin and Michael Collins made the journey to the Moon and Armstrong became the first human to walk on its surface. The achievement effectively marked the end of the Space Race (it was very expensive to keep going), but it accomplished its goal of getting humankind off our planet and reaching for the stars.

A Brief History of Missions to the Moon: the Luna and Apollo Programmes

Although Apollo 11 was the first mission to get people to the Moon, a number of other missions also came under the banner of Apollo. And, as shown in the table below, the Soviet Luna programme was getting objects to the Moon well before and significantly after the Apollo missions.

Mission	Launch date	Remarks
Luna 1	2 January 1959	Flyby: missed intended impact with the Moon; first spacecraft ever to fall into orbit around the Sun.
Luna 2	12 September 1959	Impactor: impacted the Moon, becoming the first human-made object to reach the surface.

Mission	Launch date	Remarks
Luna 3	4 October 1959	Flyby: first photographs taken of the far side of the Moon.
Luna 4	2 April 1963	Problem with upper stage rocket and mission failed.
Luna 5	9 May 1965	Soft lander: was to be first soft lander on the Moon but impacted instead due to retrorocket failure.
Luna 6	8 June 1965	Soft lander: failed mid-course correction caused probe to fly off beyond the Moon.
Luna 7	4 October 1965	Soft lander: failure due to premature retrorocket fire.
Luna 8	3 December 1965	Soft lander: although one of the lander's airbags was pierced (sending it into a spin) and its retrorockets fired late, it was able to conduct most of the mission, including testing its stellar guidance system.
Luna 9	31 January 1966	Soft lander: after 11 tries, Luna 9 became the first spacecraft to make a soft landing on the Moon's surface. It sent back pictures of the lunar landscape and confirmed that the lunar regolith could support the weight of a spacecraft.

Mission	Launch date	Remarks
Luna 10	31 March 1966	Orbiter: Luna 10 was the first artificial satellite of the Moon. On board instrumentation included a meteorite detector, a magnetic field measurement device and a gamma-ray detector.
Luna 11	24 August 1966	Orbiter: designed to study the lunar surface by measuring X-ray and gamma-ray emissions, and streams of meteorites and solar wind. Mission completed 227 lunar orbits before batteries failed.
Luna 12	22 October 1966	Orbiter: designed to photograph the lunar surface. Mission completed 602 lunar orbits before losing contact with Earth.
Luna 13	21 December 1966	Lunar lander: the probe soft landed on the lunar surface near the Ocean of Storms, and a lunar soil tester probe beamed back images of lunar surface.
Apollo 1	27 January 1967	Crew died before launch due to a spark igniting the high-pressure, pure oxygen atmosphere during a rehearsal.

Mission	Launch date	Remarks
Apollo 2–6		Technology test missions with no people on board.
Luna 14	7 April 1968	Orbiter: mission lasted eight days and was designed to test communication system for future lunar landing probes.
Apollo 7	11 October 1968	Earth orbital test flight, which included first live TV broadcast from American spacecraft.
Apollo 8	21 December 1968	First lunar orbits: ten orbits in 20 hours. First manned flight of Saturn V.
Apollo 9	3 March 1969	First manned flight test of Lunar Module (LM): tested propulsion, rendezvous and docking. An EVA (Extra Vehicular Activity, or spacewalk) tested the portable life support system.
Apollo 10	18 May 1969	'Dress rehearsal' for lunar landing. LM descended to 15.6km from lunar surface.
Luna 15	13 July 1969	After completing 52 lunar orbits, the space probe began descent to take samples but crash-landed on 21 July near Sea of Crises. This took place while Apollo 11 astronauts were on the lunar surface.

Mission	Launch date	Remarks
Apollo 11	16 July 1969	First manned landing in the Sea of Tranquillity; featured one surface EVA.
Apollo 12	14 November 1969	Precise Moon landing – the first ever – in the Ocean of Storms. Two surface EVAs were conducted.
Apollo 13	11 April 1970	Intended landing on Fra Mauro cancelled after Service Module (SM) oxygen tank exploded. LM used for safe crew return. First time that a S-IVB stage was crashed on the Moon for seismic tests.
Luna 16	12 September 1969	Sample return: first robotic probe to land on lunar surface and return sample. Probe landed near Sea of Fertility. Returned 101g of lunar regolith.
Luna 17	10 November 1970	Lander and rover: first deployment of *Lunokhod* rover, first remote-controlled rover to be used on another celestial body. Travelled over 10km on lunar surface near Sea of Rains. Initially designed to work for three lunar days (three months) but mission extended and completed 11 lunar days.

Mission	Launch date	Remarks
Apollo 14	31 January 1971	Successful landing at Fra Mauro. Colour TV images broadcast from lunar surface for first time. Two surface EVAs. Science experiments conducted in space for first time.
Apollo 15	26 July 1971	Successful landing at Hadley–Apennine. Lunar stay lasted three days. Lunar Roving Vehicle used for the first time. Conducted three lunar surface EVAs plus one deep space EVA (to reclaim orbital camera film from the Service Module).
Luna 18	7 September 1971	Was to be a sample return mission but after completing 54 lunar orbits the soft landing went wrong and the spacecraft impacted the lunar surface near Sea of Fertility.
Luna 19	28 September 1971	Orbiter: studied lunar gravitational field, discovering areas of mass concentrations (mascons).

Mission	Launch date	Remarks
Luna 20	14 February 1972	Sample return: soft landing in the mountainous region called the Terra Apollonius near Sea of Fertility. Returned a 30g sample of lunar regolith to the Earth.
Apollo 16	16 April 1972	Landed in Descartes Highlands. Conducted three lunar EVAs and one deep space EVA.
Apollo 17	7 December 1972	Landed at Taurus–Littrow. The first time a launch took place at night. Conducted three lunar EVAs and one deep space EVA.
Luna 21	8 January 1973	Lander and rover: successfully landed on lunar surface and deployed *Lunokhod 2*. The rover travelled nearly 40km over lunar surface.
Luna 22	29 May 1974	Orbiter: mission objectives to image lunar surface and to measure Moon's magnetic and gravitational field.
Luna 24	9 August 1976	Sample return: last of the Luna missions, landed in Sea of Crises and returned 170g of lunar regolith.

CURRENT MISSIONS TO THE MOON: THE LAST TEN YEARS

So where are we now, given that the Apollo programme ended nearly half a century ago? Over the last ten years there has been a resurgence in missions to the Moon but they have not only been conducted by the usual suspects. China, India and Japan now seem to be moving centre stage in the exploration of our local partner.

China has perhaps the most ambitious plans and, over the last decade, it has been planning and testing a series of space missions that are due to culminate in a double sample-return mission by the end of the decade.

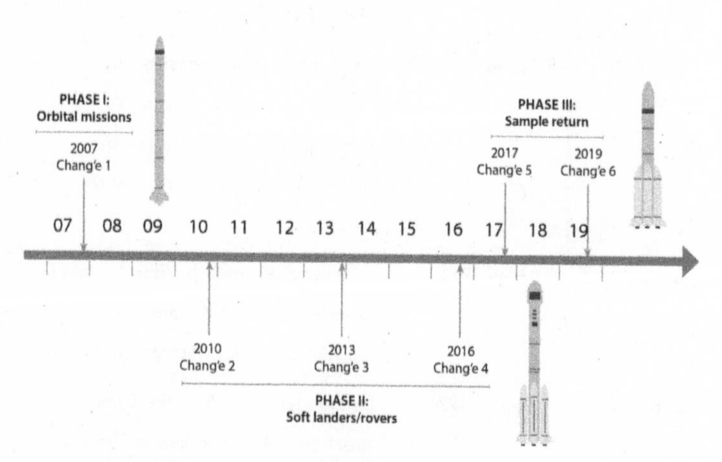

Summary of China's recent and proposed future lunar missions.

Below are some of the scientific highlights of these recent missions.

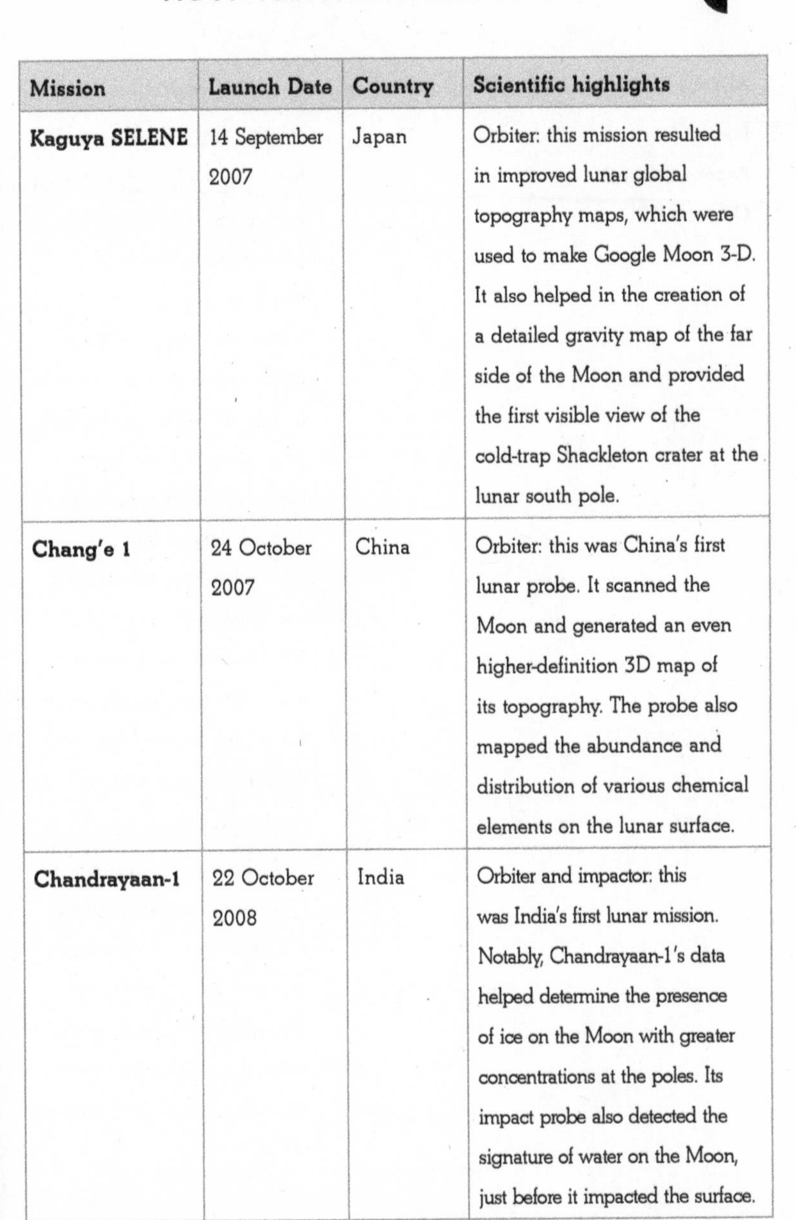

Mission	Launch Date	Country	Scientific highlights
Kaguya SELENE	14 September 2007	Japan	Orbiter: this mission resulted in improved lunar global topography maps, which were used to make Google Moon 3-D. It also helped in the creation of a detailed gravity map of the far side of the Moon and provided the first visible view of the cold-trap Shackleton crater at the lunar south pole.
Chang'e 1	24 October 2007	China	Orbiter: this was China's first lunar probe. It scanned the Moon and generated an even higher-definition 3D map of its topography. The probe also mapped the abundance and distribution of various chemical elements on the lunar surface.
Chandrayaan-1	22 October 2008	India	Orbiter and impactor: this was India's first lunar mission. Notably, Chandrayaan-1's data helped determine the presence of ice on the Moon with greater concentrations at the poles. Its impact probe also detected the signature of water on the Moon, just before it impacted the surface.

Mission	Launch Date	Country	Scientific highlights
Lunar Reconnaissance Orbiter (LRO)	18 June 2009	US	Orbiter: NASA's eye in the lunar sky, the LRO, is designed to map the lunar surface in great detail to enable the planning of future missions to the Moon's surface. It has taken a series of high-resolution images of the Apollo landing sites, showing kit and vehicle tracks on the lunar surface. A global map of the lunar surface has been taken with a resolution of 100m per pixel. The LRO also contained a microchip with the names of 1.6 million people who wanted their names on the Moon.
LCROSS	18 June 2009	US	Orbiter and impactor: LCROSS was launched with LRO and was designed as a follow-up of the detection of water by Chandrayaan-1. The spacecraft experienced some difficulties but was able to confirm the presence of water.

Mission	**Launch Date**	**Country**	**Scientific highlights**
Chang'e 2	1 October 2010	China	Orbiter: the next stage of the Chinese lunar mission, the scientific objectives of Chang'e-2 were to achieve overall moon imaging with 7m resolution and to create a distribution map of a number of elements on the Moon's surface. At the end of its extended mission on the Moon, the spacecraft was propelled into space and now sits over 200 million km away from Earth and hopes to get to 300 million km.
Gravity Recovery and Interior Laboratory (GRAIL)	10 September 2011	US	Orbiter: GRAIL A and GRAIL B, renamed Ebb and Flow, operated in a nearly circular orbit near the poles of the Moon at an altitude of about 55km. The distance between the twin probes changed slightly as they flew over areas of greater and lesser gravity such as mountains and craters, and mascon hidden beneath the lunar surface. Their results caused a re-evaluation of the way some of the Moon's larger craters were formed.

Mission	Launch Date	Country	Scientific highlights
Lunar Atmosphere and Dust Environment Explorer (LADEE)	6 September 2013	US	Orbiter: this mission was designed to analyse the Moon's very thin atmosphere. The most abundant elements in the lunar atmosphere (called an exosphere as it is so thin) were found to be helium, argon and neon. The helium and neon were found to be supplied by the solar wind.
Chang'e 3	2 December 2013	China	Lander and rover: Chang'e 3 was the first lander/rover mission of the Chinese Lunar Exploration Programme. As well as gathering high-resolution images to locate a suitable landing site, Chang'e 3 was able to perform a successful soft landing on the Moon and deploy a rover.

Mission	Launch Date	Country	Scientific highlights
Chang'e 5-T1 (Test Vehicle)	23 October 2014	China	Flyby and return: this was the test vehicle for the Chang'e 5 sample-return mission. The vehicle was put in a brief Earth parking orbit before transferring into lunar orbit. It then went around the Moon and pictures of the Moon and Earth were taken throughout the mission. The vehicle touched down in Inner Mongolia and was returned to Beijing for analysis. The Chang'e 5 mission is due to be launched in 2019, delayed after the failure of its launch vehicle. Meanwhile, Chang'e 4 is scheduled to land in late 2018.

MOON FUTURE:

WHAT LIES AHEAD?

WHEN WILL WE RETURN TO THE MOON?

When speaking of human missions to the Moon, we usually quote the epic words of Neil Armstrong as he stepped out onto the lunar surface for the first time. 'That's one small step for a man, one giant leap for mankind.' But to examine the path of our future on the Moon over the next few decades, I think that it may be more appropriate to start with some words from our last human mission there.

In 1972, Apollo 17 was the last manned mission to the Moon's surface (unfortunately no women have been to the surface yet), and just before its commander Gene Cernan climbed on board the spaceship and sealed the hatch before departure he uttered these words:

> 'I'd like to just say what I hope history will record: that America's challenge of today has forged man's destiny of tomorrow ... we leave as we came and, God willing, as we shall return, with peace and hope for all mankind.'

Powerful words – but some 50 years later we are still hoping to achieve what his remarks inspired. In more recent years, Cernan, who died in 2017, went on record to ask:

> 'When are we going back?'

Until recently, it did not seem likely that we ever would. Over the last few years, the established agencies have lacked focus, ambition and, above all, cash – but now the rivalry to return to the Moon is heating up.

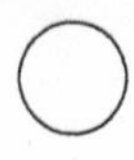

Why is this? In a word: competition. Looking at the last ten years of lunar exploration, the gene pool is getting increasingly varied, with more and more countries realising the benefits the Moon has to offer. Not wanting to be on the back foot to exploit them, they are funding new missions of exploration.

Remote sensing is giving us a tantalising glimpse of the resources available on the surface, resources that the Apollo and Luna missions only hinted at. With only 13 soft-landing sites, all close to the equator, and all on the near side of the Moon, it is clear now that they did little more than scratch the surface, both metaphorically and literally, and there is evidently much more to be explored. Combined data from the Apollo missions and the latest remote missions mean that a targeted approach to future exploration could yield impressive results.

But as well as countries, private organisations are reaching for the Moon too. Some of the most well covered by the media have been inspired by the Google Lunar X Prize. This was a competition launched in 2007 as an inducement to encourage private organisations to develop technologies that will enable cheaper access to the Moon. To win the prize, a company would have to get a spacecraft to the Moon's surface, for it to travel 500m (1,640 feet) over the lunar terrain, and for it to also be able to transmit high-definition video and images back to Earth.

In February 2011, 32 teams had registered for the challenge but by January 2017 only 5 remained. After a number of deadline extensions, in January 2018 the prize was officially listed as unclaimed as no one had managed to achieve the three goals. However, the impact of the competition has been widespread, including the first commercial space companies formed in India, Malaysia, Israel and Hungary, as well as large capital sums, totalling $300 million, being raised for the projects across the world. Many jobs were created as a result.

Other well-publicised space companies such as SpaceX, Virgin Galactic and Bigelow Aerospace are looking to commercialise travel to the Moon but, to date, fulfilling Cernan's last words on the lunar surface remain an aspiration rather than a reality.

There is a bigger question, though, that lies behind these efforts and that is simply – why? Why go to all the effort of returning humans to the Moon anyway? The answer can be broken down into two main areas: science conducted on the Moon, and commerce involving the Moon's resources and location, including its opportunities for space tourism.

THE FUTURE OF SCIENCE ON THE MOON

An expert in this area and a former colleague of mine is Ian Crawford, a professor of planetary science and astrobiology who now works at Birkbeck, University of London. Human space travel has long been a passion of his and he has sat on a number of committees, looking at the logistics of getting us out there. He believes, as he puts it, that 'science is not, and is never likely to be, the sole motivation for human space activities. Nevertheless ... planetary science stands to be a major beneficiary of human space exploration, especially as regards the geological exploration of the Moon and Mars.'

But what science can be done on our moon that will justify some of the set-up costs? Well, the Moon has some unique characteristics that are virtually impossible to recreate here on Earth.

Minimal Atmosphere

Here on Earth, astronomy is rather impeded by the Earth's atmosphere. Due to its turbulent nature, telescopes larger than 4m (157 inches) in diameter experience so much movement of the image under observation that only limited scientific study is possible. If you go outside at night and look at the sky, the twinkling of stars is caused by the turbulent atmosphere. In the past, telescopes were launched into space to overcome this problem. However, space telescopes are very expensive so we are limited in the number and size that we are able to deploy. In the last 40 years, new technology called adaptive optics has been developed for ground-based telescopes, which can literally remove the twinkle from stars and other objects under observation (see page 194). They work well but a telescope on the Moon would give us the best of both worlds. We could get the clarity of a space telescope but with some of the infrastructure that would support a ground-based telescope.

An example of the challenges incurred by a space telescope is Hubble. After it had been launched on 24 April 1990, a flaw discovered in one of the optical components rendered it of limited capability. To fix it, a space shuttle was deployed with astronauts on board to fit a corrector lens, at a cost of around $85 million to the US taxpayer.

DE-TWINKLING THE STARS

It sounds so romantic, life under the twinkling stars, but to astronomers that twinkling is a pain we could do without, as it limits what we can observe and resolve. Although the twinkling of the stars seems part and parcel of their nature, it is actually caused by a phenomenon local to us: turbulence in the atmosphere.

Over the last 40 years, technology has been available to 'de-twinkle' them for clearer observations from the ground, using a method called adaptive optics. It involves using a bright 'reference star' that sits close to the object under observation. This serves as a reference point with which to monitor and measure the extent of the atmospheric disruption. Having measured what the atmosphere is doing, sophisticated computer programs calculate (within a few milliseconds) the inverse of the disruption. This information is then sent to a deformable mirror that sits in the light path through the telescope. Its shape is changed to neutralise the turbulence and produce a stable image, by correcting the 'wobble' introduced by the Earth's atmosphere. This is all done in close to real time.

There are hundreds of thousands of these natural reference stars in the night skies, but there are still only enough to allow this method to cover about 1 per cent of the sky. So scientists came up with a way of creating artificial guide stars, if there was no suitable natural one nearby.

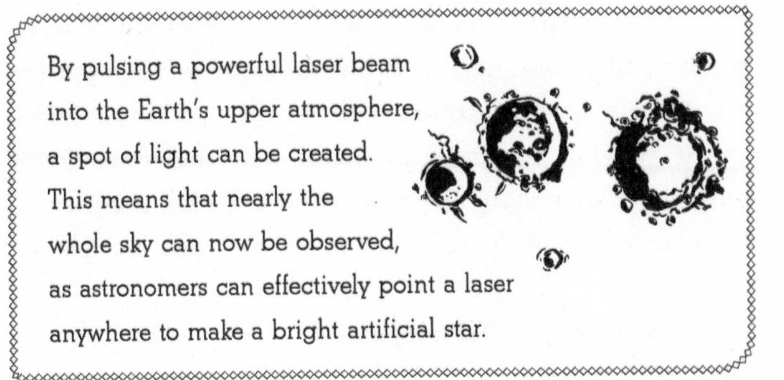

By pulsing a powerful laser beam into the Earth's upper atmosphere, a spot of light can be created. This means that nearly the whole sky can now be observed, as astronomers can effectively point a laser anywhere to make a bright artificial star.

Radio Silence

The Moon would be a very good location for radio astronomy – this is the astronomy that studies objects in space that emit radio waves. Although this can be done from the Earth's surface, the faint signals that are received from space can be swamped by radio signals produced locally on Earth. However, a radio telescope on the far side of the Moon would be shielded from Earth's copious radio noise. Another challenge with Earth-based radio telescopes is that only certain radio frequencies emitted by objects out in space can be detected. The radio window consists of frequencies that lie in the range from about 5MHz to 30GHz. At the low-frequency end of this window, radio signals generated from different astronomical sources are reflected off the Earth's ionosphere* back into space. At the upper end of

* The ionosphere is a layer of the Earth's atmosphere that sits around 75 to 1000 km above the Earth's surface. In this region molecules lose their electrons due to solar radiation (this makes them ions) creating a thin layer of electrons. The ionosphere is used for radio communications on Earth as it has the ability to reflect radio waves generated on Earth to other parts of the planet.

the frequency spectrum, radio waves are absorbed by water vapour and carbon dioxide in the atmosphere. By contrast, the Moon's tenuous atmosphere means that observations across the full spectrum of radio frequencies would be possible. Opening up a window into these unobserved frequencies will likely lead to many exciting new discoveries.

Low Gravity: Bigger Telescopes

No matter what sort of light astronomers – amateur or professional – are trying to gather (X-rays, visible light, radio waves, ultraviolet), the one thing that we are all after is a bigger telescope. Larger telescopes have the ability to gather more light, and more light enables us to see more clearly with better resolution.

Here on planet Earth, construction of large telescopes is expensive. The European-Extremely Large Telescope (E-ELT) is an optical telescope with a primary (main) mirror that is 39m (128 feet) in diameter. (And yes, that really is its name, following in the footsteps of its smaller cousins, the VLTs, i.e. the Very Large Telescopes, four 8m [26 feet] telescopes that sit in the Atacama Desert in Chile where the ELT is also under construction). It is being built by a consortium of European countries at an estimated cost of $1.2 billion.

If we were to build the ELT telescope on the Moon just now it would, of course, cost a lot more money. However, if in the future the infrastructure already existed on a Moon base then the price of this telescope could go down significantly. In a gravitational pull of only 16.7 per cent of that of Earth, less of a support structure would be needed so much bigger telescopes could be considered. Any lifting would also be much easier and the use of elements found in the lunar regolith

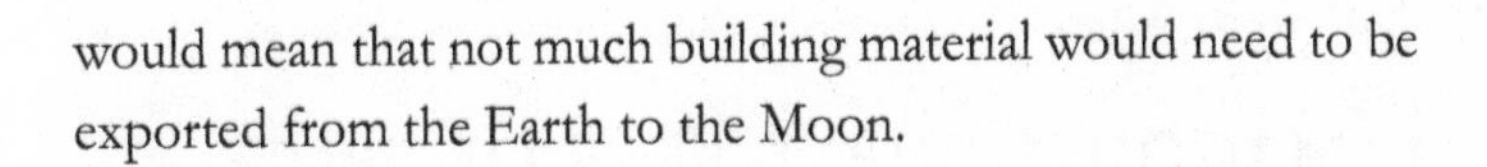

would mean that not much building material would need to be exported from the Earth to the Moon.

Light-Free Craters

As we discussed in the section on observing the Moon (see page 136), light pollution can be a real problem when observing space from Earth. Radiation from the Sun, be it in the form of light or radio waves, can swamp out the tiny signal we are trying to detect. We now know that on the Moon there are craters that do not receive any light; parts of them are in constant shadow. These would be very desirable locations for an optical telescope. Such telescopes here on Earth can only work for around half of the day: while the Sun is in the sky it produces so much light pollution that we optical astronomers can only work after the Sun has set. A telescope in a light-free crater, by contrast, could operate 24/7.

A Time Capsule in the Sky

So far, all of the science we have spoken of falls in the category of things that are generally already done on Earth but could be done better on the Moon. But there is a feature of the Moon, that has no parallel on Earth. The Moon with its minimal erosion and tenuous atmosphere, keeps a virtually unblemished record of our solar system's past history. We know that it has been bombarded with debris left behind after the formation of the solar system, and this makes it a unique time capsule for us to study our origins and to get an understanding of our local place in the solar system. Lunar samples that have been returned to Earth have given us a tantalising glimpse, but getting actual human geologists onto the lunar surface, digging into the past, will reveal more than I think we can ever imagine.

MAGGIE'S THREE ERAS OF SPACE AND HOW COMMERCIALISATION WILL GET US OUT THERE

For a long time I have had a theory that we have had three eras of space since the Space Age began on 4 October 1957 with the launch of Sputnik 1.

1. **Confrontation**: the space era was born out of a "need" to lob intercontinental ballistic missiles over great distances, using technology developed during the Second World War. So space exploration emerged from a very dark past. We morphed from this into the space race, where space was a vehicle to show global dominance.

2. The second era is **Collaboration**: this is the period in which I have spent my career as a space scientist. This period has seen such gems as the formation of the European Space Agency, the International Space Station and one of my favourite moments in history, when in July 1975 a Russian Soyuz and an American Apollo capsule docked together and exchanged the first international handshake in space. But as Gene Cernan stated, this era has perhaps lacked some of the vision and most of all cash that the confrontation era had.

3. The third era of space, and the one I am most interested in, is **Commercialisation**: to my mind this is what gets industry moving and I think that it is commerce that will take us on the next step in our journey into space.

THE FUTURE OF COMMERCE ON THE MOON

So the Moon is truly a playground of discovery for scientists but, as Ian Crawford said (see page 192), science is not enough in itself to drive future missions to the Moon's surface. To really get our feet off the ground, a viable lucrative reason to travel out there is needed. In this section, we look at some of the commercial reasons that might pave the way for future missions, from harnessing the Moon's natural resources and power, to providing the ultimate destination for a holiday . . .

Resources

Before we get into looking at the possible commercial use of lunar resources, it is good to remind ourselves how we already know what we know about the composition of the Moon. As discussed before, there are three main sources of data. The first is from the Soviet Luna and the US Apollo missions, which together returned just under 400kg of lunar samples which were then analysed by scientists around the world, some of which are at the Lyndon B. Johnson Space Center in Houston. The second source is the lunar meteorites that are released from the Moon due to impacts and which occasionally fall to Earth. The third and more recent source is high-tech remote sensing.

I have felt very privileged to have spent time at the Johnson Space Center and to get my hands on a small amount of the original moonrock samples. Back in the 1970s, geologists were able to analyse these samples and work out that the lunar surface is rich in minerals.

So what do these samples tell us? Well, the composition of the rocks analysed was dependent on the location that the lunar rocks came from; for example, whether they were from the maria or the highlands. Rocks obtained from sites in the lunar maria showed large traces of metals, with around 15 per cent aluminium oxide (Al_2O_3), 12 per cent calcium oxide (lime), 14 per cent iron oxide (rust), 9 per cent magnesia (MgO), 4 per cent titanium dioxide (TiO_2) and just over 0.5 per cent sodium oxide (Na_2O).

Samples from lunar highlands were similar in composition but with different proportions, with 24 per cent aluminium oxide, 16 per cent lime, 6 per cent iron oxide, 7.5 per cent magnesia, and over 0.5 per cent titanium dioxide and sodium oxide.

One of the things to come out of these studies was that lunar rocks contain large amounts of oxidised minerals. This means that large amounts of oxygen are present, even if it is locked into chemical compounds. However, experiments can be performed to release this oxygen, which could be used to provide astronauts with breathable air or to make water and even rocket fuel.

Concentrations of rare earth metals were also found on the lunar surface. These are a group of 17 chemically similar elements, some of which are crucial to the manufacture of high-tech products. Despite their name, most are abundant on Earth but can be hazardous to extract.

Because of their usefulness in the manufacture of the high-tech goods, rare earth metals are becoming increasingly important to the global economy. As 90 per cent of the world's current reserves are controlled by one country, China, there is great interest in an alternative source of these metals from the Moon.

Water

At the time of initial analysis, no water was discovered in the samples from the Apollo missions and the Moon was thought to be bone dry. This was a disappointing find and further scrutiny did not seem important. But recently, the samples have been analysed again, now with twenty-first-century equipment, and have been found to contain hitherto undetected amounts of hydroxyl, a compound produced when magma containing water cools down. It seems that the Moon does have water inside it, rather large amounts in fact.

The extent of the water is still up for debate but some think that there could be enough water inside the Moon to fill the Great Lakes two and a half times over. Or, looked at another way, if you took all of the lunar water and put it on the surface, it would make a 1m-thick layer covering the entire Moon.

But even today, the Moon's surface has been found to have significant amounts of water contained within the lunar regolith at trace concentrations of 10 to 1,000 parts per million, with greater concentrations found in the permanently shadowed areas – the cold traps – of the northern and southern polar regions. Like oxygen, this water would also be valuable as a source of rocket fuel, breathable air and drinking water for future astronauts.

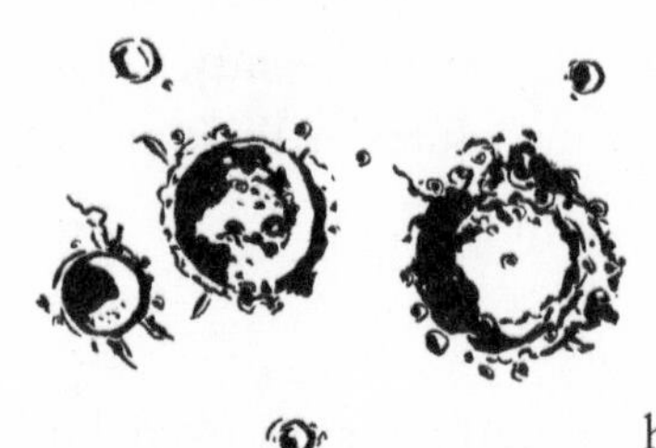

Since this analysis, multiple remote-sensing missions have not only detected water on the lunar surface but may have revealed evidence of where it has come from.

Chandrayaan-1, India's first mission to the Moon, had both an orbiter and an impactor. In November 2008, the impactor was released and made a 25-minute descent towards the lunar surface, from which it found evidence of water in the Moon's thin atmosphere. In November 2009, the US LCROSS mission had similar findings around the southern polar region. Its impactor kicked up material from the lunar surface, which analysis showed contained water. In March 2010, Chandrayaan-1's orbiter also discovered more than 40 permanently darkened craters near the Moon's north pole that are believed to contain a few million litres of ice. And in 2012, surveys conducted by the US Lunar Reconnaissance Orbiter (LRO) revealed that ice makes up to 22 per cent of the material on the floor of the Shackleton crater (located in the southern polar region). *The Moon is wet!*

But where did the water come from? It seems that this water could have been delivered via two different methods. Firstly, there's the good old regular bombardment theory again, but this time the Moon is being hit by water-bearing comets, asteroids and meteoroids. The second theory is that the water is being produced locally by the hydrogen ions of solar wind combining with oxygen-bearing minerals found in the lunar regolith to make H_2O.

Helium-3

As well as water, another resource that is spoken of as one of the most valuable commodities on the Moon is the isotope helium-3.

The helium-3 atom is made as a by-product of the fusion reactions* that take place in the heart of the Sun. This process is what powers the Sun and releases huge amounts of energy. It occurs when temperatures and pressures are reached that allow atomic nuclei to fuse together to make new elements and release energy.

Helium-3 leaves the Sun's surface and is carried away into space by the solar wind through the solar system and eventually out beyond. But as we have discovered before, the solar wind can crash into objects that get in its way, depositing materials on them: for example, an object just like the Moon. Scientists haven't been able to find any reliable sources of helium-3 here on Earth, but it seems that significant quantities may be available on the lunar surface.

Very little helium-3 is used on Earth today, but scientists across the world are trying to set up fusion reactors that can exploit the huge potential energy source that fusion would provide. The ideal fuel for these reactors is helium-3, which could potentially provide a clean, non-radioactive fuel for the

---

* Fusion reactions are governed by Einstein's special relativity equation, $E=mc^2$. The E stands for the energy released, m is the mass consumed and c is the speed of light. The speed of light, c, is equal to approximately 300 million metres per second (m/s), so for this equation c gets multiplied by itself to give nearly 90 million billion m^2/s^2 . From this equation we can see that a very little bit of mass can generate huge amounts of energy, and it is this principle that is used in an atomic bomb.

nuclear fusion reactors of the future. However, helium-3 exists on the Moon trapped within the lunar rock. The only way to extract it would be to strip-mine the Moon, a plan that would not be without controversy (see page 205).

Lunar Power

The lunar landscape providing helium-3 for future fusion reactors on Earth may not be the only way that the Moon can supply power to us all.

It is thought that by 2050 the human population will top 9.7 billion people, compared with the 7.3 billion that we have today. Oil reserves will have run down and there will be a huge demand for new energy sources. One solution to the energy problem is to build more solar farms – but rather than locating them on Earth, one plan is to build them on a vast scale on the Moon and to beam the energy back to Earth via microwaves.

Without an atmosphere, the Moon is an ideal location for solar panels, better than any of the sun-drenched wildernesses on Earth. And with some lunar locations receiving sunlight close to 24 hours a day, solar panels on the Moon seem to make a lot of sense. It may even be possible to make the solar panels from lunar dust, with robotic machines scooping up the regolith and manufacturing solar panels on site to line the lunar surface.

AN ETHICAL DILEMMA

From an ethical point of view, should we be considering these options? Do we really want to cover large chunks of the Moon with solar panels – panels that would inevitably be visible from Earth? As for the environmental damage of a strip-mining operation, is that what we want for the Moon?

Are we humans going back to the Moon simply to take what we can, at any cost? Or should we treat the Moon as a wilderness – a sanctuary to be cherished and preserved, much like Antarctica?

Given the likelihood that we will run out of conventional fossil fuels relatively soon, these are not easy questions to answer. But my inner 'lunatic' instincts tell me that we should tread very carefully before altering the Moon in any significant way.

Space Tourism

The Moon has so many potential resources that we could use in the future but one that I have not mentioned until now is the idea of space tourism. To date, just a handful of people have paid to travel into space. And those that have paid, have paid a *lot*. But when will we all get a chance to travel out there – be it sub-orbital, in a low Earth orbit, to the Moon or even out further to Mars?

The idea of space tourism always makes me think of international flights in the 1950s. These journeys were reserved for the great and the good. The image of a famous, glamorous starlet, chihuahua in hand, getting on board a flight to somewhere exotic springs to mind. International travel used to be something that people aspired to, but within a period of just 70 years we have gone from that to the EasyJet culture of today, with relatively cheap flights for everyone. (A prime example of this is my daughter Lauren: by the age of four she had clocked up over 100 flights, whereas I made my first flight when I was 14.)

What drove the transformation to flights for all? Essentially it was two things: demand by the public and a response from commerce, which saw an opportunity and drove the price down with technological developments to make it more accessible yet turn a good profit. And it is not just international flights that have had this treatment. Look at the price drop of computer hard drives since the 1980s. Or mobile phones – initially the size of a brick and only for wealthy businesspeople, whereas now we all have slimline versions that do everything except make the tea.

My hope is that space tourism will go the same way. As many of us desire to get out there, technology will be accelerated to meet the demand and cheap, safe flights to the Moon will become accessible. But for this to happen we need to develop an infrastructure on the lunar surface. This has been discussed for many years, but when will it actually happen?

PROPOSED MOON BASES

The idea of a Moon base has been around for many years. Talk of putting an infrastructure on the Moon often seems like a good stopgap solution for presidents looking to inspire their people; let's face it, it worked for Kennedy. But not many are willing to put a country's money where its president's mouth is.

After the Second World War and at the start of the Space Race, many proposals for Moon bases were put forward. During this period, in the 1950s, there was a certain optimism in the air; anything seemed possible and a lunar colony must have seemed like the logical progression from the early days of lunar exploration.

An example of this is episode 47 of *The Sky at Night* (the number of episodes now runs into the 750s), first shown in September 1963. In this programme Arthur C. Clarke appeared as a guest, discussing ideas for proposed lunar bases with Patrick Moore. To put this into context, this was a period just six years after the birth of the space era and still six years before the Moon landings occurred.

The drawings shown on the programme of the proposed Moon base had a distinctively 1950s sci-fi feel to them. Some of the images dated as far back as 1948 and had been devised by a chap called Ray Smith. They showed pressure domes that contained the living inhabitants and tubular farms surrounding the central base, which would be used for the farming of both food and breathable air in an attempt to make the base as self-sufficient as possible.

As an interesting aside, Clarke at the time felt that the Russians would be the first to land people on the Moon and return them to Earth – he predicted it would happen in 1968. Despite the US spending $10 million per day on the Moon project at the time, he thought the Americans would not be capable of a manned landing on the Moon until the early 1970s. Clarke also predicted that there would be a Martian base within the century.

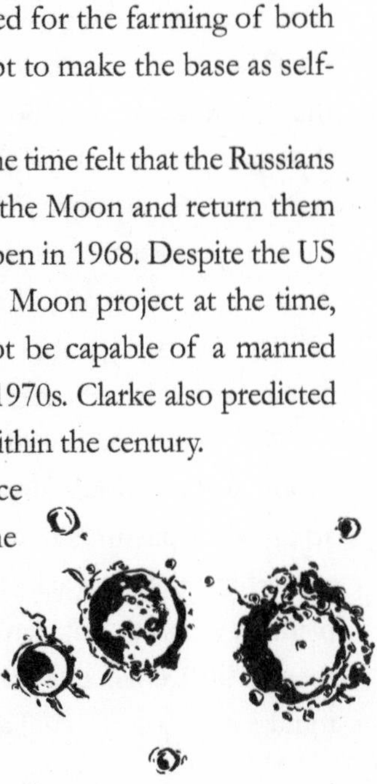

During this time, many other space agencies issued proposals for the creation of a lunar base and, just as in the plans discussed by Clarke, all the proposed bases were designed to be as self-sufficient as possible, utilising local resources for nearly all of their needs. However, as the US bill racked up for the Apollo programme and interest in the Moon landings waned in the seventies and eighties, all ideas of such bases were largely abandoned. But, in recent decades, detailed proposals are once again being put forward.

During the George W. Bush administration (2001–9), NASA was commissioned to create a design for a 'lunar outpost' as part of its Vision for Space Exploration (2004). It came up with a plan for the construction of a Moon base to be carried out between 2019 and 2024. In a similar way to Clarke's proposals, the plan was to use lunar regolith wherever possible to produce the key components needed for the base.

These plans never really got off the ground and the whole programme was cancelled by the Obama administration. Instead, NASA was asked to come up with a plan for a Mars base.

This sort of oscillation shows one of the difficulties with the planning of a Moon base. There are two schools of thought: one is that we return to the Moon, establish a base and learn the lessons of living in a hostile environment, but close to home. Once the base has been established, then it can act as a gateway to other locations in the solar system. We could launch our space probes from the low-gravity, hence cheaper, Moon rather than the high-gravity, more expensive Earth.

The second school of thought is that we have 'done' the Moon and we should use the lessons learned from this to go straight for pastures new, such as Mars, rather than reviving something that was achieved 50 years ago. To my mind both plans have merit but, while the debate continues, I feel that there is little chance for either one to get enough momentum to take off.

PROS AND CONS OF ESTABLISHING A BASE ON MOON VS MARS VS THE MYSTERY THIRD OPTION

Moon

Pros	Cons
Relatively close, no launch windows needed.	Very inhospitable, crazy temperatures, no atmosphere, no protection.
Interesting discoveries to make.	Lunar regolith dangerous to handle.
Resources on the Moon could allow base to be built from local materials.	Lung disease from regolith.
Good staging post.	Very low gravity is bad for humans in the long term.
Water found on the Moon.	
Gateway to rest of solar system.	

Mars

Pros	Cons
Interesting new science to be done there.	Much further away.
Stronger gravity than Moon.	Astronauts would be isolated, less support.
Water found on Mars.	No protection on Martian surface.
Possibility of life on Mars to investigate.	More infrastructure needed.

Mystery Third Option: the Venusian Atmosphere, 50km above surface of Venus, living in a dirigible

Pros	Cons (if reliant on surface)
Manageable temperatures of 0–50°C.	Hell on surface: temperatures that would melt lead.
Pressure just right to feel very at home.	Sulphuric acid rain.
	Crushing pressure.

In 2014, a group of scientists from NASA, Harvard and the Lunar X Prize held a workshop looking at low-cost options for returning to the Moon. One of the outcomes of the meeting was to describe how a settlement could be built there by 2022 for just $10 billion. This was possible due to the advances in various technologies such as global launch capabilities and development of novel industries such as autonomous robots and 3D printing.

In December 2016, an international symposium entitled 'Moon 2020–2030: A New Era of Human and Robotic Exploration', hosted by the European Space Agency, looked at creating a roadmap for future human and robotic missions to the Moon. The meeting was well attended with representatives from all the major space agencies, including ASI (Italy), CNES (France), CNSA (China), CSA (Canada), CSIRO (Australia), DLR (Germany), ESA (European Space Agency), ISRO (India), JAXA (Japan), KARI (Republic of Korea), NASA (USA), NSAU (Ukraine), Roscosmos (Russia) and UKSA (United Kingdom).

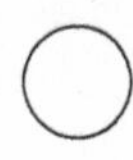

The meeting investigated the agencies' desire to create an international lunar base using robotic workers that could possibly be controlled remotely by human operators, along with 3D printing techniques and the utilisation of *in situ* resources.

Many other space agencies are also developing plans for lunar bases. The Russian space agency (Roscosmos) issued plans to build a lunar base by the 2020s, and the China National Space Agency (CNSA) has proposed to build such a base in a similar timeframe, thanks to the success of its Chang'e lunar space probe programme.

As well as the government organisations mentioned above, many private companies are looking towards the Moon as the basis of future development and revenue. The company SpaceX, however, which has been a true pioneer in the private space industry, is not participating in the Moon vs Mars debate. It has abstained from this because it is making plans to travel to *both* destinations and more. With the company's new launch vehicle, the BFR (its working title, standing for Big Falcon Rocket), it is planning a single vehicle that would be able to reach destinations like Mars and the Moon, but which would also be able to cover local transport needs on Earth, cutting journey time by a significant margin. The benefit of this single-vehicle approach is that a commercially viable system that generates income here on Earth can support the less-funded or unfunded excursions to the Moon and planets. The hope is that by concentrating on one system, research will be more focused and that the limited funds available can be used to cover all of the transport needs of a space-faring world.

To my mind, the engineering challenge of this is a daunting prospect: taking on any one of these goals would be hard but amalgamating them into one vehicle will be extremely difficult and may slow down progress. However, few commercial space companies have had the success that SpaceX has had in the past few years, so I guess if anyone can, SpaceX can.

Route	Distance	Time Taken via Commercial Airline	Time Taken via BFR
Los Angeles to New York	3,983km	5 hours, 25 minutes	25 minutes
Bangkok to Dubai	4,909km	6 hours, 25 minutes	27 minutes
Tokyo to Singapore	5,350km	7 hours, 10 minutes	28 minutes
London to New York	5,555km	7 hours, 55 minutes	29 minutes
New York to Paris	5,849km	7 hours, 20 minutes	30 minutes
Sydney to Singapore	6,288km	8 hours, 20 minutes	31 minutes
Los Angeles to London	8,781km	10 hours, 30 minutes	32 minutes

An overview of the (proposed) SpaceX BFR capabilities on Earth.

President Donald Trump signed a directive in December 2017 that included an initiative to send astronauts to the Moon – and eventually to Mars. The timescales seemed a bit vague and the pledge may well just be hot air but, at this moment in time, it looks as if we have a greater will in the space community – national, international and private – for a lunar base to happen. However, is it a case of our having been here before? Do we yet have the critical mass needed and, if we do, what would a lunar Moon base be like?

LIVING ON THE MOON

The Moon will be, by any standards, a hostile environment for the first lunar settlers. Any structures built there will have to withstand its huge solar temperature ranges and the lack of atmospheric pressure, so any structures will need to be resilient to puncture. Moonquakes would be a real danger to inhabitants, especially as they have such long duration, so structurally any base will need to withstand this threat. Another potential problem is the threat of meteors on the Moon. Ideally any housing will have a protective barrier to minimise impact.

So what will lunar housing look like? Well, any shelter will need to protect inhabitants from radiation – one idea is to have an inflatable habitat which, once in place, is covered with lunar regolith. If possible, it should be built using the local regolith, as the costs of transporting materials to the Moon would be immense. And scientists are currently working on robotic 3D printers to do the actual constructing as it is likely that humans will not be there at the beginning.

The aim is for food to be grown locally in lunar greenhouses, using LED lighting and hydroponics (growing plants without soil). Water would be mined from the dark lunar craters or the oxidants in the regolith. It's water that is the key component to making a lunar settlement successful – it is essential for life, but also (in the form of hydrogen and oxygen) it's the key to transportation in space as rocket fuel.

WHO OWNS THE MOON?

All this talk of lunar bases and access to the solar system in the future gets me fired up and keen to get started as soon as possible. It is the fulfilment of my childhood dreams coming to life. But before I get carried away, there is a fly in the ointment that could cause all these dreams to come tumbling down.

We have established that for a Moon base to work, it needs to be driven by commercial interests, but then the question remains: who owns the Moon?

Most people's initial response would be all of us (despite the US flag being planted on it) but legally it is a quagmire.

There is some legislation on this subject already in place. It is called the Outer Space Treaty and was signed by many key nations in 1967. But its genesis had a long and chequered history.

After the Second World War, talks were undertaken to see if outer space could be preserved for peaceful purposes.

These talks began in the late 1950s at the newly formed United Nations. The US and its allies submitted proposals in 1957 on reserving space exclusively for 'peaceful and scientific purposes', but the Soviet Union rejected these proposals. It seems like an odd stance now, but at the time the Soviets were on the cusp of launching Sputnik, the world's first satellite. As these initial forays into space were designed to find the most efficient route for launching an intercontinental ballistic missile, and the Soviets were in a prime position to start testing, it may be less surprising that they were not particularly keen to agree to the proposals.

In 1963, the UN General Assembly approved two resolutions on outer space that later became the basis for the Outer Space Treaty. One called on countries to refrain from stationing Weapons of Mass Destruction (WMD) in outer space and the other set out the legal principles of outer space exploration, which stipulated that all countries have the right to freely explore and use space.

The United States and the Soviet Union submitted separate draft outer space treaties to the UN General Assembly in June 1966. A mutually agreed treaty text was eventually signed by the Soviet Union, US and UK on 27 January 1967, and entered into force on 10 October that year.

The key details of the treaty are summed up rather well by Arms Control Association, who issued a fact sheet on the subject*. Firstly, states commit not to:

- place in orbit around the Earth or other celestial bodies any nuclear weapons or objects carrying WMD;

* www.armscontrol.org/print/2555

- install WMD on celestial bodies or station WMD in outer space in any other manner;
- establish military bases or installations, test any type of weapons, or conduct military exercises on the Moon and other celestial bodies.

Other treaty provisions underscore that space is no single country's domain and that all countries have a right to explore it. These provisions state that:

- space should be accessible to all countries and can be freely and scientifically investigated;
- space and celestial bodies are exempt from national claims of ownership;
- countries are to avoid contaminating and harming space or celestial bodies;
- countries exploring space are responsible and liable for any damage their activities may cause;
- space exploration is to be guided by 'principles of cooperation and mutual assistance', such as obliging astronauts to provide aid to one another if needed.

The Outer Space Treaty has now been in place for over 50 years and is considered to be rather long in the tooth by many. The challenge now is how can we cover the eventualities of what future technologies and opportunities will bring.

Recently, President Trump has suggested the development of a 'space force' as a new branch of US military to maintain their dominance in space. The idea is unlikely to take off but it shows why treaties like this need to be enforced.

Interestingly, a Moon treaty was proposed by the UN in 1979. In summary, this more recent treaty applies not only to the Moon but all other celestial bodies excluding Earth. It gives jurisdiction to the international community and requires that all activities must be conducted in accordance with international law. It states that the Moon should be used for the benefit of all states and all people of the international community. It also seeks to avoid the Moon becoming a source of international conflict.

It all sounds like good, useful stuff, but although it received the required number of ratifications to come into force, none of the major space-faring players signed up, so at this point in time it remains largely irrelevant. No one is in contravention of the agreement, but in the future, as more activities occur on the Moon and Mars, then the agreement may be redrafted and hopefully receive a better reception.

There is one area on Earth that is akin to the Moon and has legislation in place to control its use, and that place is Antarctica. Antarctica has no government and no country owns it, yet a treaty exists (the Antarctic Treaty) to govern its use. This was signed in 1959 by a number of countries. Although interesting mineral resources are thought to reside on the continent, to date no one country has tried to exploit this and it is mainly used as a location for research bases.

MOON, MARS AND BEYOND?

The Moon vs Mars argument will, I think, continue for many years to come, though it seems like the idea of using the Moon as a staging post to the solar system and maybe beyond is currently looking like the most likely solution. It seems that if we can get a Moon base established, many other opportunities will become available. The idea of a lunar gateway seems to be gaining traction but, as always, the greatest challenge seems to be how to fund such an enterprise. Let's see what the future brings.

One thing is certain: whatever space technology we develop, and whenever we make our next return to the Moon, let us hope, as Gene Cernan suggested back in 1972, that we will do it in the spirit of peace and hope for all mankind.

CONCLUSION:

LOOKING OUTWARDS

AND SO our journey comes to an end. We have gone on a voyage that has explored the physical, mental and emotional impact of the Moon on all of us. But, of course, what has been discussed here is just a drop in one of the lunar oceans.

If, like me, you would like to release your inner lunatic, you may actually want to travel there physically rather than just admire it from afar.

It reminds me of a dream that I once had as a teenager, one of those dreams that stays with you all of your life because of its powerful emotional charge.

I was in a big, fluffy bathrobe having just had a wonderfully relaxing bath. I stepped out into a warm, comfortable room and in it were all the people I loved. I picked up my baby, who was lying contentedly in a crib, and carried her over to a large window. Looking out, I saw the deep blackness of space. It looked wonderful, better than I have ever seen it, and there, suspended in that wondrous view, was the partially lit Earth.

It gave me a feeling of deep serenity and when I woke I wanted to return again and again to that moment.

I hope that in the very near future we will all be able to live in space and create scenes like the one I experienced in my dream – but how do we all get there?

Well, I think that commercialisation may succeed in the job, but rather than just relying on this avenue, I hope that the countries of the world can collaborate to make the vision happen.

Our ancestors had long hoped to travel to the Moon, Mars and beyond. We have taken that first step and maybe the Moon will also be the stepping stone that enables us to reach the stars.

ACKNOWLEDGEMENTS

TO ALL the many people who have helped me create this book.

To BBC Books who encouraged and enabled me to write about a lifelong passion of mine, the Moon.

To Laura, who pushed me and empowered me to create a book that far exceeded my expectations.

To Lindsay, who was able to steer my dyslexic rambling sentences into something far more coherent – similar to what was going on in my head, but was hard to translate into the written word.

To Helen and the team at the BBC who enabled me to present the documentary *Do We Really Need the Moon?*, some material of which is included here.

To my long suffering agent, Vicki and the many friends and colleagues who understood when emails remained unreturned, responses were lacking and I was incommunicado due to being in 'the throes of writing' 24/7.

To Martin and Lori, who survived the terrors of my writing mode, which often led to chaos around the house and a terrifying example of the dreaded 'Zombie Mummy' syndrome.

To each and every one of you and the many others not mentioned here, I give my boundless thanks.

And finally, to the Moon itself, my friend and companion, who I continue to have an ongoing affair with, that makes me distraught at times of the new Moon and during cloudy skies when I cannot see you to make contact. Thank you for just being there and all you do for us lesser mortals. (Apologies, a little of my inner lunatic showing again.)

INDEX

Page references in *italics* indicate images.